The Handbook
of Environmental Chemistry

Volume 4   Part C

Edited by O. Hutzinger

Air Pollution

With contributions by
C. Gries, F. W. Lipfert, M. Lippmann,
T. H. Nash

With 46 Figures and 10 Tables

Springer-Verlag
Berlin Heidelberg GmbH

Professor Dr. Otto Hutzinger

University of Bayreuth
Chair of Ecological Chemistry and Geochemistry
P.O. Box 101251, W-8580 Bayreuth,
Federal Republic of Germany

ISBN 978-3-662-14997-3

Library of Congress Cataloging-in-Publication Data
(Revised for vol. C)
Air pollution. (The Handbook of environmental chemistry; v. 4, pt A–C)
Includes bibliographies and indexes.
1. Air Pollution. 2. Environmental chemistry. I. Dop, H. van (Han van), 1944– . II. Series: Handbook
of environmental chemistry; v. 4, etc. QD31.H335 vol. 4, etc. 540 s [628.5′3] 86-219604 [TD1]

ISBN 978-3-662-14997-3   ISBN 978-3-540-47343-5 (eBook)
DOI 10.1007/978-3-540-47343-5

© Springer-Verlag Berlin Heidelberg 1991
Originally published by Springer-Verlag Berlin Heidelberg New York in 1991
Softcover reprint of the hardcover 1st edition 1991

Typesetting: Macmillan India Ltd., Bangalore-25

52/3020-543210–Printed on acid-free paper

Preface

Environmental Chemistry is a relatively young science. Interest in this subject, however, is growing very rapidly and, although no agreement has been reached as yet about the exact content and limits of this interdisciplinary subject, there appears to be increasing interest in seeing environmental topics which are based on chemistry embodied in this subject. One of the first objectives of Environmental Chemistry must be the study of the environment and of natural chemical processes which occur in the environment. A major purpose of this series on Environmental Chemistry, therefore, is to present a reasonably uniform view of various aspects of the chemistry of the environment and chemical reactions occurring in the environment.

The industrial activities of man have given a new dimension to Environmental Chemistry. We have now synthesized and described over five million chemical compounds and chemical industry produces about one hundred and fifty million tons of synthetic chemicals annually. We ship billions of tons of oil per year and through mining operations and other geophysical modifications, large quantities of inorganic and organic materials are released from their natural deposits. Cities and metropolitan areas of up to 15 million inhabitants produce large quantities of waste in relatively small and confined areas. Much of the chemical products and waste products of modern society are released into the environment either during production, storage, transport, use or ultimate disposal. These released materials participate in natural cycles and reactions and frequently lead to interference and disturbance of natural systems.

Environmental Chemistry is concerned with *reactions in the environment*. It is about distribution and equilibria between environmental compartments. It is about reactions, pathways, thermodynamics and kinetics. An important purpose of this Handbook is to aid understanding of the basic distribution and chemical reaction processes which occur in the environment.

Laws regulating toxic substances in various countries are designed to assess and control risk of chemicals to man and his environment. Science can contribute in two areas to this assessment: firstly in the area of toxicology and secondly in the area of chemical exposure. The available concentration ("environmental exposure concentration") depends on the fate of chemical compounds in the environment and thus their distribution and reaction behaviour in the environment. One very important contribution of Environmental Chemistry to the above mentioned toxic substances laws is to develop laboratory test methods, or mathematical correlations and models that predict the environmental fate of new chemical compounds.

The third purpose of this Handbook is to help in the basic understanding and development of such test methods and models.

The last explicit purpose of the handbook is to present, in a concise form, the most important properties relating to environmental chemistry and hazard assessment for the most important series of chemical compounds.

At the moment three volumes of the Handbook are planned. Volume 1 deals with the natural environment and the biogeochemical cycles therein, including some background information such as energetics and ecology. Volume 2 is concerned with reactions and processes in the environment and deals with physical factors such as transport and adsorption, and chemical, photochemical and biochemical reactions in the environment, as well as some aspects of pharmaco-kinetics and metabolism within organisms. Volume 3 deals with anthropogenic compounds, their chemical backgrounds, production methods and information about their use, their environmental behaviour, analytical methodology and some important aspects of their toxic effects. The material for volumes 1, 2, and 3 was more than could easily be fitted into a single volume, and for this reason, as well as for the purpose of rapid publication of available manuscripts, all three volumes are published as a volume series (e.g. Vol. 1; A, B, C). Publisher and editor hope to keep the material of the volumes 1 to 3 up to date and to extend coverage in the subject areas by publishing further parts in the future. Readers are encouraged to offer suggestions and advice as to future editions of "The Handbook of Experimental Chemistry".

Most chapters in the Handbook are written to a fairly advanced level and should be of interest to the graduate student and practising scientist. I also hope that the subject matter treated will be of interest to people outside chemistry and to scientists in industry as well as government and regulatory bodies. It would be very satisfying for me to see the books used as a basis for developing graduate courses on Environmental Chemistry.

Due to the breadth of the subject matter, it was not easy to edit this Handbook. Specialists had to be found in quite different areas of science who were willing to contribute a chapter within the prescribed schedule. It is with great satisfaction that I thank all authors for their understanding and for devoting their time to this effort. Special thanks are due to the Springer publishing house and finally I would like to thank my family, students and colleagues for being so patient with me during several critical phases of preparation for the Handbook, and also to some colleagues and the secretaries for their technical help.

I consider it a privilege to see my chosen subject grow. My interest in Environmental Chemistry dates back to my early college days in Vienna. I received significant impulses during my postdoctoral period at the University of California and my interest slowly developed during my time with the National Research Council of Canada, before I was able to devote my full time to Environmental Chemistry in Amsterdam. I hope this Handbook will help deepen the interest of other scientists in this subject.

                                                                                                    Otto Hutzinger

This preface was written in 1980. Since then publisher and editor have agreed to expand the Handbook by two new open-ended volume series: Air Pollution and Water Pollution. These broad topics could not be fitted easily into the headings of the first three volumes.

All five volume series will be integrated through the choice of topics covered and by a system of cross referencing.

The outline of the Handbook is thus as follows:

1. The Natural Environment and the Biogeochemical Cycles,
2. Reactions and Processes,
3. Anthropogenic Compounds,
4. Air Pollution,
5. Water Pollution.

Bayreuth, June 1991   Otto Hutzinger

Contents

List of Contributors

C. Gries
Arizona State University
Dept. of Botany/Microbiology
Tempe, AZ 85287-1601
USA

Prof. Dr. Morton Lippmann
Institute of Environmental Medicine
Anthony J. Lanza Research Laboratories
Long Meadow Road
Tuxedo, N.Y. 10987
USA

Prof. T. H. Nash
Arizona State University
Dept. of Botany/Microbiology
Tempe, AZ 85287-1601
USA

Dr. Frederick W. Lipfert
23 Carll Court
Northport, N.Y. 11768
USA

Lichens as Indicators of Air Pollution

T.H. Nash III and C. Gries

Department of Botany, Arizona State University, Tempe, AZ 85287-1601, USA

Summary

Lichens are well known as sensitive indicators of air pollution, particularly for sulfur dioxide. In part, this is related to their unique biology. Evidence supporting this assertion goes back well over 100 years and is based on extensive field and laboratory studies. In general, these studies reinforce each other, but for oxidants the data are not entirely consistent, and consequently require further work. At the least, lichens appear to be less sensitive to oxidants than vascular plants. Acid precipitation effects are closely related to SO_2 effects. The mechanistic basis for SO_2 effects is briefly reviewed. The extreme sensitivity of lichens to SO_2 is partially related to their ability to absorb more SO_2 for a given concentration than typical vascular plants. The use of lichens as long-term integrators of elemental deposition patterns is well established, but their use for monitoring dry deposition has only recently been recognized. Air pollutants adversely impact not only growth, reproductive potential, and morphology, but also a wide variety of physiological processes, which also becomes reflected in ultrastructural changes. The impact of organic pollutants on lichens is largely undocumented and is a prime area for future work, even though much work remains to be accomplished with the traditionally recognized air pollutants.

Introduction

Lichens are widely recognized as useful indicators of air pollution [1–3]. This statement is supported by extensive field studies in regions where air pollutants are known to occur and by experimental studies with known air pollutants. The initial statement is, of course, a generalization that requires cautious interpretation and limited extrapolation. For example, all lichens are not equally sensitive to all air pollutants. Rather different lichen species exhibit differential sensitivity to specific air pollutants. As a consequence, one can frequently relate variation in lichen communities, as occurrence of certain species or the occurrence of morphological anomalies, to specific air pollutants and in some cases to the concentration of that air pollutant. However, there are certainly constraints on the degree to which this can be done. Factors other than air pollutants will also influence patterns within lichen communities. In addition, lichens can not be used to monitor short-term (hourly or shorter) atmospheric concentrations. For example, the development of visual injury symptoms in lichens is not sufficiently specific that one can document the probable occurrence of an ozone fumigation, as one can with a sensitive variety of tobacco [4]. To the contrary, lichens, as long-lived perennial organisms, are useful integrators of air pollutant patterns over years or, in some cases, over weeks or months.

On the other hand, lichens are frequently useful monitors of deposition patterns in cases where no specific injury to the lichen occurs. They have been used extensively to document the long-term (over years) deposition of metal ions around point sources and in urban areas [5]. In addition, recent data [6, 7] document that lichens can be effectively used to monitor biweekly accumulation of ions by dry deposition.

Finally, the generality of the statement that lichens are sensitive indicators of air pollution is constrained by our knowledge of what the air pollutants are. Certainly, sulfur dioxide, hydrogen fluoride, nitrogen oxides, and ozone are well-recognized air pollutants today, but specific identification of ozone as an air pollutant goes back only three and half decades to the pioneering work of Haagen-Smit [8]. In the future we will doubtlessly recognize currently unknown air pollutants. For example, organics, a large class of compounds released by both natural and industrial sources may well include a number of air pollutants not recognized today.

The Biology of Lichens and its Relation to Air Pollutant Sensitivity

Our knowledge of lichen biology is well summarized in introductory texts by Hale [9] and Hawksworth and Hill [10]. In addition, a number of more advanced texts

[11] and reviews [12–16] are available. Much of the information developed in this chapter comes from these sources and, as more general knowledge, are not referenced further.

Although lichens are classified as fungi, they are actually symbiotic organisms that are autotrophic because of the presence of at least one photobiont, either an alga and/or cyanobacterium. In most cases the photobiont accounts for less than 10% of biomass of the lichen, but it is the principal carbon source for the fungus (the mycobiont). The symbiosis is frequently referred to as an excellent example of mutualism, but more recently Ahmadjian & Jacobs [17] have argued that it should be viewed as controlled parasitism. Although photobiont growth is limited, lichenization has provided one opportunity for utilizing terrestrial environments, that would probably be uninhabitable for each symbiont alone. Certainly, the symbiosis is a successful one as almost 15,000 species are recognized from all 7 continents.

Lichens occur in most of the world's ecosystems, either as epiphytes growing on trees or as separate plants on exposed rock and/or soil. They play important ecological roles [18, 19] as interceptors of wet and dry deposition, as nitrogen fixers in the case of cyanobacteria-containing species, as carbon fixers in ecosystems where they are abundant, in the weathering of rocks, and the retention of soils in arid regions. In North America they are the dominant plant group in approximately 8% of the terrestrial ecosystems, most of which are found at northern latitudes where they form extensive carpets approximately 1 dm thick [20]. In size, lichens vary from some pendulous forms that hang up to several meters from trees, to crustose forms that may occupy less than a square cm. Morphologically they vary from large shrubby or pendulous forms (fruticose species), to small to large "leaf-like" forms (foliose species) that grow over their substrates, to crustose forms that are imbedded within their substrates. The anatomy of crustose species may be relatively simple compared to foliose or fruticose taxa, but in almost all cases the photobiont is found in a layer beneath a fungal cortex. Although some lichens increase their biomass by 20–25% per year [21], many other species grow more slowly. Most species live for decades or hundreds of years and a few are estimated to live for 1000 to 2000 years.

In comparison with vascular plants, lichens differ in a number of important ways. Lichens have no vascular system for conducting water or nutrients. As a consequence, they have no mechanism for efficiently taking up water and nutrients from the soil, as do vascular plants with their roots. Furthermore, lichens have no well-developed means for retaining water, such as the leaf cuticle and stomate system provided by vascular plant leaves. Thus, the water status of a lichen varies passively with changes in the water status of the environment, whereas most vascular plants are continuously hydrated except under drought conditions. The result is that lichens are frequently dehydrated, but they are also capable of absorbing dew, fog, and even water vapor from non-saturated air under conditions of high humidities. The lack of a root system also means that lichens must use atmospheric sources for much of their nutrient supply, although blowing soil particles are certainly trapped within their thalli.

From an air pollution perspective, these attributes of lichens are important in a number of ways. Alteration of the symbiotic balance between the photobiont and mycobiont may readily lead to a break-down of the association. As long-lived, perennial organisms, lichens are exposed to air pollutants all year. Unlike many vascular plants they have no deciduous parts and hence cannot avoid pollutant exposure in this way. The lack of stomata and cuticle means that aerosols may be absorbed over the entire thallus surface. Thus, air pollutant gases are assumed to readily diffuse down to the photobiont layer. Although dehydration allows lichens to survive dry periods, it also concentrates solutions to the point that toxic concentrations may occur. Dependence on atmospheric sources of nutrients usually implies survival on relatively dilute solutions, that become concentrated solutions in the presence of some air pollutants. Lichen decline may result not only from the occurrence of toxic concentrations, but also from favorable nutrient supplies that favor one symbiont over the other.

Historical Perspective

Independent observations in England, Munich, and Paris in the 1800s documented that lichens were disappearing from urban areas [22]. The extensive use of coal for home heating and industrial applications led to clouds of smoke over cities and, as a consequence, it was inferred that air pollution might cause of their disappearance. By the early 1900s this "city" effect was a widely recognized phenomenon in Europe and many workers followed Sernander's [23, 24] classic studies in Stockholm where he recognized a *lichen desert* zone in the city center where almost all lichens had disappeared, a *struggle zone* where many species were impoverished but some survived well, and a *normal zone* where species were apparently unaffected. Such zonation patterns around cities are now recognized on all man-inhabited continents. Mapping studies [25] and other quantitative techniques [26, 27] have been extensively developed for identifying species and community patterns in relation to air pollution sources.

Early transplantation studies from normal areas into city centers demonstrated that many lichens could not survive in city centers, but they did not identify specific causes. The process of urbanization affects lichen habitats in a variety of complex ways and, consequently, it is unlikely that air pollution was the sole cause of the lichen deserts. Thus, it was also important that over the past 50 years studies of lichens around isolated point sources of pollution have also documented marked impoverishment of lichen communities as one approached the source [28]. Such studies provided further field evidence that lichens were sensitive to such air pollutants as SO_2, HF, and to some metals [29].

Initially air pollution was thought of as coal soot, but by the 1930s the colorless gas, sulfur dioxide, became widely recognized as a phytotoxic agent. By the mid-1970s experimental studies had established that many lichens were sensitive to SO_2 fumigations at concentrations known to occur in the field [30–32]. Other experimental studies provided documentation of fluoride [33–35] and zinc [36] effects.

With increasing industrialization, dispersion of air pollutants has become widespread. Concurrently, a decline of lichen floras for whole regions has been noted

[37–39]. Some species, such as *Lobaria pulmonaria*, have now disappeared over wide areas [40], and where relic stands exist, they exhibit distinct injury. On the other hand, one apparently pollution-tolerant species, *Lecanora conizaeoides*, a species collected for the first time in the late 1800s, has expanded its range throughout Europe and is probably the most common species occurring there on bark today [3].

Sulfur Dioxide: A Well-Established Case Study

By 1970 the correlation of high sulfur emissions with declines in lichen communities was sufficiently obvious that Hawksworth and Rose [41] were able to develop a semiquantitative scale for English forests relating the occurrence of

Table 1. Zone scale for the estimation of mean winter sulfur dioxide levels in England and Wales using lichens on trees with moderately acid bark. After [3].

Zone	SO_2 Concentration $\mu g\,m^{-3}$	Epiphytes
0	?	Epiphytes absent
1	> 170	*Pleurococcus viridis* s.l. present but confined to the base
2	∼ 150	*Pleurococcus viridis* s.l. extends up the trunk; *Lecanora conizaeoides* present but confined to the base
3	∼ 125	*Lecanora conizaeoides* extends up the trunk; *Lepraria incana* becomes frequent on the bases
4	∼ 70	*Hypogymnia physodes* and/or *Parmelia saxatilis*, or *P. sulcata* appear on the bases but do not extend up the trunks. *Lecidea scalaris*, *Lecanora expallens*, and *Chaenotheca ferruginea* often present
5	∼ 60	*Hypogymnia physodes* or *P. saxatalis* extends up the trunk to 2.5 m or more; *P. glabratula*, *P. subrudecta*, *Parmeliopsis ambigua*, and *Lecanora chlarotera* appear; *Calicium virde*, *Lepraria candelaris* and *Pertusaria amara* may occur; *Ramalina farinacea* and *Evernia prunastri* if present, largely confined to the bases; *Platismatia glauca* may be present on horizontal areas
6	∼ 50	*P. caperata* present at least on the base; rich in species of *Pertusaria* (e.g., *P. albescens*, *P. hymenea*) and *Parmelia* (e.g., *P. revoluta* (except in NE), *P. tiliacea*, *P. exasperatula* (in N)); *Graphis elegans* appearing; *Pseudevernia furfuracea* and *Alectoria fuscescens* present in upland areas
7	∼ 40	*Parmelia caperata*, *P. revoluta* (except in NE), *P. tiliacea*, *P. exasperatula* (in N) extend up the trunk; *Usnea subfloridana*, *Pertusaria hemisphaerica*, *Rinodina roboris* (in S), and *Arthonia impolita* (in E) appear
8	∼ 35	*Usnea ceratina*, *Parmelia perlata*, or *P. reticulata* (S and W) appear; *Rinodina roboris* extends up the trunk (in S); *Normandina pulchella* and *U. rubiginea* (in S) usually present
9	∼ 30	*Lobaria pulmonaria*, *L. amplissima*, *Pachyphiale cornea*, *Dimerella lutea*, or *Usnea florida* present; if these absent, crustose flora well developed with often more than 25 species on larger well lit trees
10	∼ pure	*L. amplissima*, *L. scrobiculata*, *Sticta limbata*, *Pannaria* spp., *Usnea articulata*, *U. filipendulla*, or *Teloschistes flavicans* present to locally abundant

approximately 50 lichen species on moderately acidic bark substrates to ranges of mean winter SO_2 measurements (Table 1). A second scale was also developed with over 30 species occurring on basic or nutrient-enriched trees [41] and modifications were developed for other European regions. Recently, it has been suggested that the scale more accurately reflects annual SO_2 measurements and the exact magnitude of the ambient concentrations is questioned based on work in Ireland and Sweden [2]. Nevertheless, the original scale, which itself was based on the pioneering work of Gilbert [42], has proved extremely useful as an elaborate hypothesis of putative SO_2 sensitivity. Because the scale was purely correlative, inferring cause and effect is an unsubstantiated exercise by itself. Experiments with SO_2 under controlled conditions are necessary.

Because it is difficult to maintain lichens under laboratory conditions for much longer than a month, direct long-term experimental fumigations with SO_2 are impractical. However, a number of short-term fumigation studies demonstrated that lichens were adversely affected by SO_2 and that differential sensitivity occurred among species (e.g., Fig. 1). This differential sensitivity allows one to rank the species and to compare this ranking with a corresponding one based on Hawksworth's and Rose's semi-quantitative scale. The Spearman rank correlation coefficient between the two ranking scales is a significant 0.72 [43]. Similarly, other correlations with other independently established experimental studies were established [43], and consequently it was concluded that the original semi-quantitative scale does in large measure reflect sensitivity to SO_2.

Furthermore, over the past two decades England has successfully implemented a program to reduce sulfur emissions. Subsequently, it has been established that a number of lichen species are reinvading areas in which they were previously absent [44–46]. Such observations strongly support the inference that the original scale

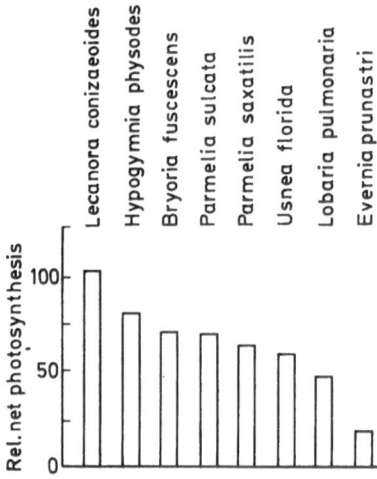

Fig. 1. Net photosynthesis (in % of control values) of nine lichen species after a 14-hour fumigation with 4.0 mg SO_2/m^3. After [32]

reflected underlying SO_2 sensitivity. Together these independent lines of evidence from both field and laboratory studies strongly support the conclusion that lichens do respond to SO_2.

It is frequently asserted that lichens are more sensitive to SO_2 than vascular plants, but this has rarely been tested directly [47]. It is difficult to imagine that the photosynthetic apparatus of lichens is more sensitive per se to SO_2 than that of soybeans or some other vascular plant. Recently, a simple explanation has emerged based on the fact that a lichen absorbs SO_2 over its whole surface, whereas vascular plants can limit gas absorption by closing their stomata [48]. For *Cladonia rangiferina*, one of the dominant arctic lichens, Grace et al. [49] carefully determined SO_2 absorption rates as a function of water content. On the basis of measured water content variability in the field [50], the seasonal SO_2 uptake potential was estimated to exceed that of vascular plants by over 100 times [48]. Greater SO_2 absorption would lead to greater physiological response and hence the appearance of enhanced sensitivity.

Of all the air pollutants investigated SO_2 is the only one for which detailed mechanistic information on effects in lichens is available [2, 51–53]. Injury mechanisms probably include (1) enzyme deactivation, (2) enzyme stimulation, (3) interaction with reactive biomolecules, and (4) formation of free radicals [52]. In solution, SO_2 forms sulfite or bisulfite, either of which can affect enzyme function by cleaving the disulfide bonds and reacting with cystine to form *S*-sulfocystine [53]. Both reactions will affect enzyme structure and hence function. Sulfur dioxide-induced deformation of chloroplasts and thylakoids is demonstrable with electron micrographs. Depending on pH, different sulfur species can act as electron donors or acceptors and thereby affect electron transport [51]. In addition, sulfite acts as a competitive inhibitor of RuDP carboxylase [54]. Also SO_2 fumigations enhance formation of free oxygen radicals [52], which may be scavenged by superoxide dismutase (SOD) and converted to hydrogen peroxide and hence affect the lipid fraction of chloroplast membranes. Several isozymes of SOD exist and Köck et al. [55] suggest several mechanisms by which the differences may account for the observed differential sensitivities of isolated lichen algae to sulfite [56]. In addition, carbohydrate transfer between photobiont and mycobiont is reduced by SO_2 fumigations [57] and apparently results in development of starch grains in the photobiont [58].

Oxidants: An Equivocal Case Study

Oxidants are a group of strongly oxidizing air pollutants formed photochemically by series of chemical reactions involving mixtures of nitrogen oxides and hydrocarbons. Ozone, which was initially identified in the Los Angeles area, is probably the most infamous oxidant and it is the most important air pollutant causing injury to agricultural crops [59]. In addition, a number of other oxidants, such as peroxyacetyl nitrate (PAN), are known to cause injury to vascular plants.

Lichen field studies have been conducted in the Los Angeles area both in coastal mountain areas [60] and in higher mountain forests [61]. Because of negligible

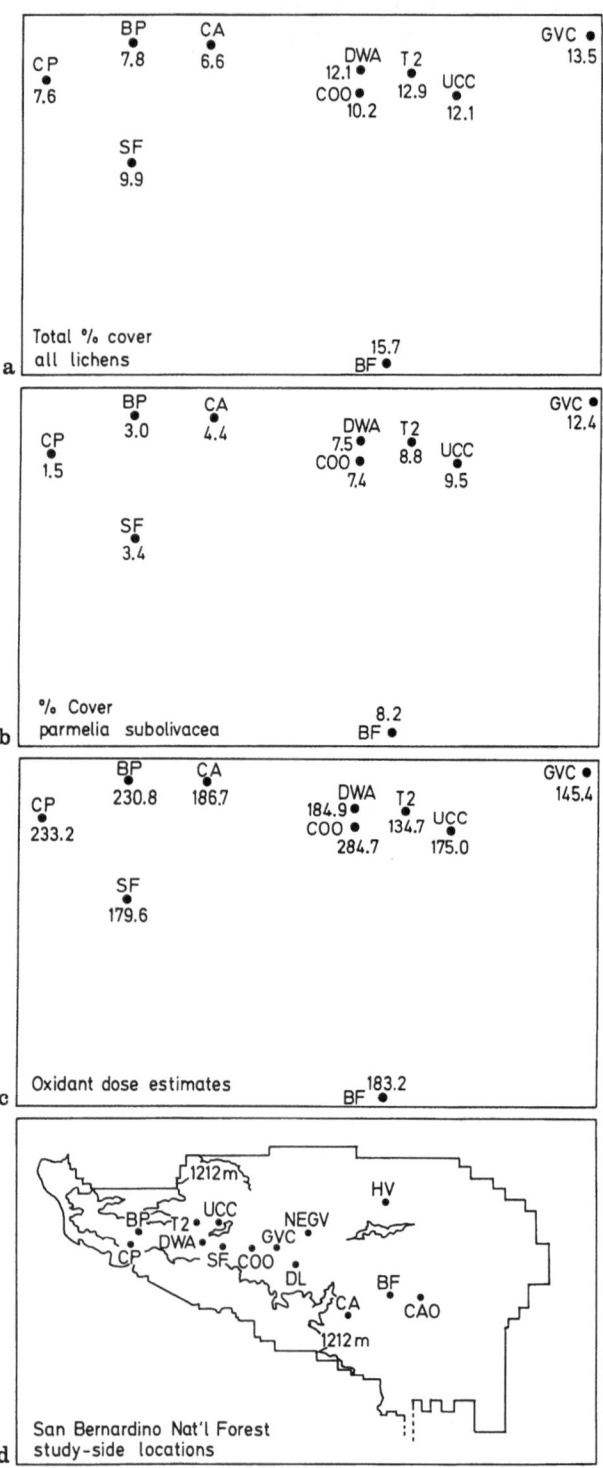

a   Total % cover all lichens

b   % Cover parmelia subolivacea

c   Oxidant dose estimates

d   San Bernardino Nat'l Forest study-side locations

Table 2. Inferred sensitivity of selected lichen species to air pollution in the Southern California mountains; according to Sigal and Nash [61].

Very sensitive	Sensitive	Moderately tolerant	Tolerant
Bryoria abbreviata	Cetraria merrillii	Hypogymnia enteromorpha	Letharia vulpina
Bryoria cf. fremontii	Collema nigrescens	Parmelia glabra	Physcia biziana
Cetraria canadensis	Leptogium californicum	P. elegantula	P. tenella
Evernia prunastri	Parmelia sulcata	P. subolivacea	Physconia grisea
Peltigera canina	P. quercina	Xanthoria polycarpa	Xanthoria fallax
P. collina	Peltigera rufescens		
P. spuria	Physcia ciliata		
Physcia sciastra	P. orbicularis		
Platismatia glauca	Polychidium albociliatum		
Ramalina farinacea	Usnea sp.		
R. menziesii			
Xanthoria candelaria			

home heating requirements, SO_2 emissions are very low and most vascular plant injury patterns are ascribed to oxidants [59]. On the basis of historical data and extensive sampling of lichens occurring on trees both in the Los Angeles area and in control areas [60–62], lichen species exhibiting differential sensitivity to the Los Angeles environs were identified (Table 2). The very sensitive species are no longer found in the Los Angeles area, although herbarium records document that many of them were abundant in the early 1900s. The sensitive species were found in vestigal amounts in protected habitats and the moderately tolerant ones were commonly present, but in reduced abundance. Frequently, the latter group exhibited marked morphological anomalies. In contrast, the tolerant species were seemingly unaffected.

Further circumstantial evidence that some of these species reflect differential sensitivity to oxidants was obtained by sampling along oxidant gradients within the highly polluted San Bernardino Mountains. The gradient reflected not only differences in oxidant dose but also a range of susceptibility symptoms exhibited by ponderosa pine. Experimental studies demonstrated that the ponderosa pine symptoms were reproducible when the trees were fumigated with ozone [59]. Along this gradient there was a marked reduction in cover of the dominant two lichens (*Letharia vulpina* and *Hypogymnia enteromorpha*) occurring on conifers [61] and also by species occurring on black oak (Fig. 2).

In support of these field data, fumigation experiments have shown photosynthetic reduction in *Hypogymnia enteromorpha*, *Flavoparmelia caperata*, and *Parmelia sulcata* with ozone exposure [63, 64] and to *Collema nigrescens*, H.

◀ **Fig. 2.** Principal component analysis ordination of 10 sites in the San Bernardino Mountains using cover data for the 5 most important lichen species occurring on *Quercus kelloggii*. Eighty percent of the variation is explained by axis one (the abscissa). **a** shows the positioning of the sites on the ordination plot with respect to total cover of all lichens; **b** the same ordination but with cover values for *Parmelia subolivacea*; **c** the same ordination, but with oxidant dose estimates (ppm-h); **d** the geographic location of the sites within San Bernardino National Forest. After [62] *enteromorpha*, and *P. sulcata* with PAN exposure [65]. However, these effects were only demonstrable when relatively high concentrations of the pollutants were used. In addition, *Ramalina menziesii*, a putatively highly sensitive species (Table 2), was unaffected by ozone [63] even when fumigated with the highest concentration ever measured in Los Angeles. Similarly, other experimental studies with ozone have not demonstrated major effects [66, 67] and field studies in an area with lower oxidant levels have not demonstrated analogous patterns in epiphytic lichens [68].

Thus, the question of whether lichens are good indicators of oxidant air pollution can only be equivocally answered. Some field data are supportive, but laboratory studies have not demonstrated the degree of sensitivity found among vascular plants. Why lichens are extremely sensitive to SO_2 but relatively insensitive to ozone remains a mystery.

Acid Rain and Fog: A Concern for the Future

Acid rain has become a major environmental issue in recent years. Both SO_2 and nitrogen oxides may react with atmospheric water to form acids. Because of widespread and high emissions of these gases from a variety of sources, such as power-generating plants, industries, and automobiles, acid deposition has become a regional phenomenon affecting extensive areas. Because CO_2 dissolved in water will form an equilibrium with bicarbonate and carbonate ions, rain is normally acid with pH 5.6. Today the pH of rains are frequently in the range of pH 4 to 5, and sometimes lower. In contrast to rain, fog is normally formed under stable atmospheric conditions and, as a consequence, may have a much higher concentration of ions. In coastal southern California, area fogs with pH 2 to 3 occur frequently and values as low as 1.69 have been reported [69]. In one location above the Los Angeles Basin, Boonpragob et al. [6] reported accumulation of 45 μmole water-leachable H^+g^{-1} over 10 weeks in transplanted material of the large fruticose lichen *Ramalina menziesii*.

Concern with acid rain and fog is not only with direct acid effects on the lichens, but also with indirect effects, such as changing substrate conditions. For example, increasing acidification of tree bark in cities [70–72], around a copper smelter [73] and recently even in remote rural areas of northern England [74] are reported. These authors were able to correlate rain pH and bark pH and to document a change in the epiphytic community to one more typical of acidic conditions. Such changes are also associated with changes in the buffer capacity of the bark [70, 73]. Kiss [72] reported disappearance of foliose thalli and a gradual decrease of generative and vegetative reproductive propagules. Around a copper smelter, LeBlanc et al. [75] found a decline in species richness from 33 species to 11 species and this was correlated with reduced pH in the lichen thalli. In more rural areas, nitrogen-fixing species, such as *Lobaria pulmonaria* and *Sticta limbata*, were virtually extinct [74]. These investigations documented a connection between rainwater pH and ambient SO_2 concentration. It is difficult to separate these two factors, particularly as interactions may occur. Puckett et al. [51], Hällgren and

Huss [76], and Türk and Wirth [77] all described enhanced toxicity of sulfur dioxide at low pH values of treatment solution or in the thalli.

Relatively few experiments have documented direct effects of acid rain or fog alone on lichens. In laboratory experiments, Sigal and Johnston [78] demonstrated differential sensitivity of gross photosynthesis among several lichen species (Fig. 3) to solutions at pH 3.3. In comparison to solutions at pH 4.3, *Umbilicaria mammulata* exhibited reduced photosynthesis of 71%; *Flavoparmelia caperata*, of 29%; and *Usnea* cf. *subfusca*, 65%. In the case of the *Umbilicaria*, the results may be seasonally dependent. Bailey and Larson [79] described a similar photosynthetic pattern for winter material whereas summer material was almost unaffected by pH. With the common reindeer lichen *Cladonia stellaris*, Lechowicz [80] reported a 27% reduction in photosynthetic capacity after treatments with pH 4 solutions high in sulfate (10.0 mg l^{-1}), but with reduced sulfate level (0.53 mg l^{-1}) only a delay in reaching maximum photosynthetic levels was found at this pH. For a 3-year field study with simulated acid rain, Lechowicz [81] only showed a significant growth reduction for treatments at pH 2.3 and a 6:1 ratio of H_2SO_4 to HNO_3. In contrast, no growth reduction was found for treatments at pH 2.3 and a 2:1 ratio of H_2SO_4 to HNO_3.

Nitrogen fixation may be more sensitive to acidity than photosynthesis. Studies with *Lobaria pulmonaria* [82], *Lobaria oregana* [83], and *Peltigera* species [84] all demonstrated a gradual decline in nitrogen fixation with decreasing pH values below pH 4. At pH 2.6, nitrogen fixation ceased altogether.

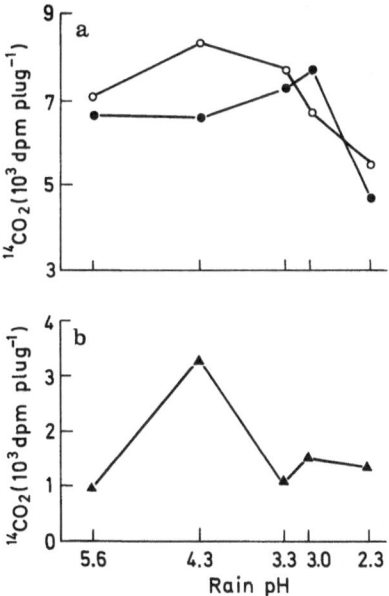

Fig. 3. Gross photosynthesis of lichens in response to simulated acidic rain: **a** data for *Flavoparmelia caperata* (●) and *Usnea* cf *subfusca* (○) are dpm per mg and **b** data for *Umbilicaria mammulata* (▲) are dpm per plug (∼ 28 mm^2). After [78]

Mechanisms accounting for the effect of acidity on lichens have not yet been elucidated. It is well known that H^+ ions will competitively displace other cations from exchange sites. In terrestrial environments, the effect of long-term exposure to acid rain may thus be a gradual decline in soil fertility, but this may have relatively little effect on lichens because of their dependence on atmospheric nutrition. On the other hand, the same principle will apply to cations held at exchange sites with the cell walls of lichens. Mineral nutrition within lichens may thus be affected.

Long-term Elemental Deposition Monitoring

Lichens are ideal biological monitors of elemental deposition for a variety of reasons [5]. Many species occupy broad geographical ranges and thus regional comparisons of elemental patterns can be made. Because they grow slowly (frequently over hundreds of years), their size changes only gradually and their morphological form does not vary seasonally. Thus, as a deposition receptor, lichens have relatively constant characteristics compared to vascular plants. Finally, lichens can accumulate elements in excess of their apparent physiological needs. These characteristics make lichens ideal for long-term elemental deposition studies and they are particularly useful in retrospective studies where pollution studies were not planned [16].

High elemental levels in lichens reflect an efficient set of accumulation mechanisms that are in part related to their dependence on atmospheric sources of elements for much of their nutrition. For example, lichens possess no roots, with their vascular system adapted to taking up nutrients from soils. Mechanisms of accumulation include particulate trapping [85, 86] in intracellular spaces that account for up to 18% of a lichen's volume [87], ion accumulation at exchange sites on cell walls [86], and intracellular uptake [88]. The latter process occurs slowly over hours following Michaelis-Menten kinetics [89], but cation uptake to exchange sites is a rapid, passive, physiochemical process occurring in a matter of minutes [90, 91]. Exchange sites are assumed to be carboxylic and hydrocarboxylic acid sites within the cell wall and calculations indicate that $6-77$ μmole g^{-1} may be held at these sites. Particulate trapping may be enhanced because lichens do not possess a surface waxy cuticle as is found in vascular plants. Around pollution sources, analysis of these trapped particles by electron probe or x-ray diffraction procedures have revealed a chemical profile almost exactly the same as the profile for particulates known to be emitted by the pollution sources [85, 92].

In addition, lichens can tolerate high elemental levels because they possess a number of mechanisms for avoiding toxicity, which has only rarely been well documented for such elements as zinc [36, 93]. Lange and Ziegler [94] proposed that tolerance mechanisms might involve (1) inherent cytoplasmic tolerance, (2) cytoplasmic immobilization and detoxification of ions by chemical combination, and (3) transport of ions to regions external to the plasmalemma and even the cell wall. There is good evidence for several different mechanisms involving the latter possibility, but relatively little evidence for the first two areas. For example, lichens occurring on high metalliferous substrates frequently sequester metals externally in oxalate crystals [95] or in complexes with some of the extracellular lichen acids [96].

Elemental data for lichens are currently available for almost 40 elements [97]. For the reasons cited above, it is not surprising that some extremely high concentrations have been reported, such as 13.5% for zinc [36], up to 5.5% for iron [94], and up to 5.9% for copper [95]. Patterns of deposition around line, point, and multiple sources are well reviewed by Puckett [5]. Around isolated point sources, exponential decreases with distance from the source has frequently been documented for the principal elements emitted [86]. Around multiple sources, deposition patterns are more complex, but use of multivariate statistical techniques in connection with multielement analyses frequently allows identification of probable sources. For example, near the city of St. John in eastern Canada, Puckett and Finegan [98] used principal components analysis to reduce variation among 7 elements and multiple sites to three principal components (Fig. 4). The first component was designated a dust or earth factor because of heavy weightings for iron, titanium, and chromium; component two, an industrial factor because of weightings for sulfur, vanadium, and nickel; and component three, a vehicle exhaust factor, because of weightings for lead.

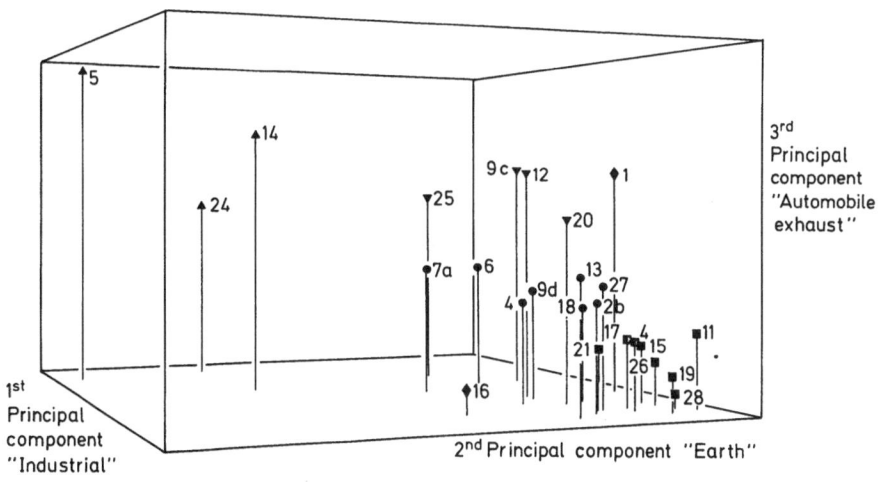

▲ Highly contaminated group   • Intermediate group for all elements

▼ Intermediate group, high lead   ■ Background group

♦ Outlier sites

Fig. 4. Lichen element content of numbered collection sites near St. John, New Brunswick, Canada, plotted on the first three obliquely rotated axes of a principal components analysis. Symbol coding of sites reflects observed element content of lichens. Labels on axes reflect authors' interpretation of atmospheric sources contributing to variation in observed element content of lichens. After [98]

Short-term Dry Deposition Monitoring

Although lichens are well known as interceptors of wet deposition in the forms of dew, fog, and rain [99–101], it has only recently been recognized that lichens are excellent monitors of dry deposition that occurs between precipitation events. With canopy lichen biomass ranging up to 700 to 3300 kg ha^{-1} in a number of forest systems [21, 102], lichens may rival leaves as interceptors of deposition. Fruticose lichens are particularly important because they add vertical structure to a canopy and, because they are highly dissected compared to leaves, these lichens will add more surface area than would be judged from biomass values alone. Quantification of dry deposition on leaves by leaching studies has been a major problem because a

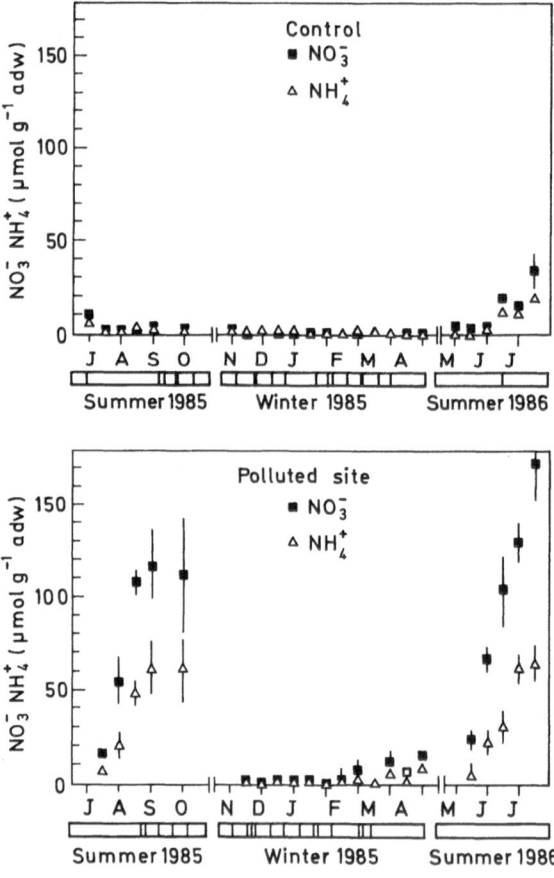

Fig. 5. Cumulative leachates of NO_3^- and NH_4^+ from the thalli of *Ramalina menziesii* at the control site (Lake Henshaw) and the polluted site (Tanbark Flat) during two-week sampling intervals from the transplant periods in summer 1985, winter 1985, and summer 1986 (n = 3 ± 1 s.e.). The three horizontal bars below each graph represent the three transplant periods with rain events marked as vertical lines. After [6]

leaf's nutrient supply is connected to the soil nutrient pool by the vascular system that extends down to the roots [103]. Thus, leaching of leaves following dry periods yields elemental data that cannot be unambigously related to dry deposition because a portion probably originates from the soil system [104]. In contrast, lichens are independent epiphytes with no direct connection to the soil pool, and hence, are more tractable monitors.

Recent transplant studies with the lichen *Ramalina menziesii* in southern California document that this lichen accumulates substantial amounts of water-leachable ions during dry periods [6, 7]. For example, approximately linear nitrate accumulation occurred over a 10-week period (Fig. 5), summer 1986 at the polluted site) during a time when measured concentrations in the atmosphere were approximately constant. Although this lichen has now disappeared from the Los Angeles area [61], it still occurs abundantly both to the north and south along the Pacific coast and at one location we have measured over 700 kg ha^{-1} of *R. menziesii* [105]. If water-leachable nitrate accumulated in such a canopy up to 171 μmol g^{-1} (Fig. 5), then leachable nitrate would be 7.5 kg ha^{-1}, a value that exceeds total yearly nitrate fluxes through many ecosystems [106].

During the same time period, other water leachable ions accumulated up to 60, 45, 25, 3, 2, and 2 μmol g^{-1} respectively for NH$_4^+$ (Fig. 5), H$^+$, Cl$^-$, F$^-$, SO$_4^{2-}$, and PO$_4^{3-}$ [6]; and up to 12, 80, 13, 7, 15, 2, 21 μmol g^{-1} respectively for Mg, K, Ca, P, Na, Fe, and Si [7]. Lesser quantities (nmol g^{-1}) of Pb, Zn, Cd, Cu, Mn, Sr, and Ba also accumulated. With the macroelements Mg, K, Ca, and P, the data need cautious interpretation because part of their origin was probably intracellular. During this summer period the lichen was severely injured [107] and disruption of cellular membranes doubtlessly occurred (see also Sect. below). However, for many of the ions the high correlation with leachable concentrations and atmospherically measured concentrations is consistent with an atmospheric origin. Future studies can address the question of origin unambiguously by quantifying the partitioning of ions among extracellular and intracellular components using procedures developed by Brown and Buck [108].

Pollutant Effects on Lichens

Using lichens as indicators of air pollution requires objective methods for evaluating the impact on the lichens. Frequently used response variables include changes in reproductive potential, gross morphology, growth, physiology, and biochemistry [52, 53, 109]. Some of the most widely used responses are measurements of membrane integrity as conductivity of leachate [110] or K$^+$ loss [111], photosynthesis and respiration [32], nitrogen fixation [76], and pigment degradation [30, 51]. Recent studies include examination of a number of other metabolic parameters as well as ultrastructure [112]. For evaluating SO$_2$ impact, Fields and St. Clair [57] compared the more commonly used responses with respect to sensitivity, cost, portability and time. Membrane permeability measured as conductivity of leachate was the most sensitive test, followed by K$^+$ loss and photosynthesis. On the other hand, nitrogen fixation, a property of less than 10% of the lichen species, is actually the most sensitive assay per se [52, 53], but has limited utility. The pigment status reveals a similar sensitivity as K^+ loss [57], but is judged less sensitive by Richardson and Nieboer [52].

Reproductive Potential

Reproductive failure is one simple mechanism by which species may disappear from an area. Because lichens are composed of at least two organisms, reproductive studies are complex and pertinent details are not fully known. Many species have vegetative modes (soredia, isidia, etc.) of reproducing the whole lichen. Others apparently reproduce by sexual means with each symbiont reproducing separately. Spore production primary through ascus development within ascocarps is the most common process in the mycobiont (a few lichen mycobionts are basidiomycetes), but reproduction of the photobiont is much more poorly documented. The process of lichenization whereby a fungal spore(s) becomes associated with the photobiont has been studied in the laboratory [17], but is undocumented in the field.

Field studies around both SO_2-dominated urban environments [113] and oxidant-dominated urban environments [61] have demonstrated that ascocarp production declines significantly in the more polluted areas. Furthermore, for lichens that do produce ascocarps in urban and industrial environments, spore germination is reduced [114, 115]. More recently, Belandria et al. [116] showed that both H_2SO_3 and NaF reduced spore germination for several lichen species, with the degree of inhibition being proportional to the apparent field sensitivity exhibited. The photobiont cell division rate is also known to be reduced in polluted areas, such as Sudbury [117].

Asexual reproduction is also altered by air pollutants. Margot [118] demonstrated substantial inhibition of algal cell division of germinating soredia when exposed to gaseous SO_2 or sulfate.

Gross Morphology

Because lichens grow slowly and have limited potential for experimental manipulation in the laboratory, gross morphological changes have not been induced under controlled fumigations. Nevertheless, a number of morphological changes have been observed in the field and these are probably related to air pollution. For example, Sigal and Nash [61] found that *Hypogymnia enteromorpha* exhibited marked variation in degree of convolution and bleaching in the mountains of southern California. The specimens shown in Fig. 6 were used as standards for degree of convolution and bleaching and randomly collected specimens from 5 mountain areas were classified according to 4-point scales for both degree of convolution and bleaching. Historical collections from the turn of the century showed very little convolution and no bleaching; whereas, the site with the best representation of lichens today (Cuyamaca State Park ca. 80 km E of San Diego) had over 60% of the samples exhibiting some degree of convolution and 40% were slightly bleached. In contrast, all samples from the San Bernardino Mountains, the range most influenced by air from Los Angeles, exhibited some degree of both convolution and bleaching. Because of the well-documented oxidant injury to vascular plants in this mountain range, it was inferred that oxidants were probably responsible for changes in morphology. Among the numerous field studies in polluted regions, there are many references to other morphological modifications of epiphytic lichens. For *Peltigera* species growing on soils high in metals, Goyal and Seaward [119] reported reduction in thallial size and rhizinal length, increased density of rhizines and branching of the veins on the lower surface and hypertrophy of the medulla.

Growth

In contrast to vascular plant studies, growth as a response variable in air pollution studies has been infrequently used with lichenological investigations and have not specifically been used with short-term fumigation studies. In part this is related to the relative slow growth rate of lichens. To obtain accurate data, very precise, tedious measurements must be made [16]. In studies along heavily travelled roads near Washington, D.C., Lawrey and Hale [120] found differential growth reduction with respect to thallus size. Compared to control areas, young, minute thalli grew much more slowly than old, mature thalli.

Recently, Schuster [121] carefully studied the juvenile development of *Usnea filipendula* and *Hypogymnia physodes* near Frankfurt (F.R. Germany) and in the laboratory. He found that attachment and development of diaspores during the first five months was largely unaffected by pollution. Subsequently, production of gelatinous material around the photobiont was reduced and this reduced the water storage capacity tremendously. This led to desiccation injury in all microhabitats except the most humid ones. In contrast, under controlled laboratory conditions, Marti [56] documented a marked reduction in diaspore germination and early growth in *Hypogymnia physodes* due to sulfite, nitrite, and particularly the pH of the substrate. Bark samples from polluted sites acidified by SO_2 were used for substrates. After fumigation with SO_2, the germination rate was less than on the bark alone.

As far as we can tell, one potential application of growth studies in lichens has not been exploited. The technique of lichenometry is well-developed for dating recent geologic events, such as glacier movement in regions where tree-rings are not available [10]. These procedures could be adapted to date probable abandonment of mine or smelter sites.

Membrane Integrity

Maintaining the integrity of membranes is a basic requirement for well-functioning cell metabolism. Membrane response may provide a sensitive assay of pollutant effect because absorbed pollutants will contact the membrane prior to entering the cell [109]. Changes in the permeability of membranes leads to loss of electrolytes

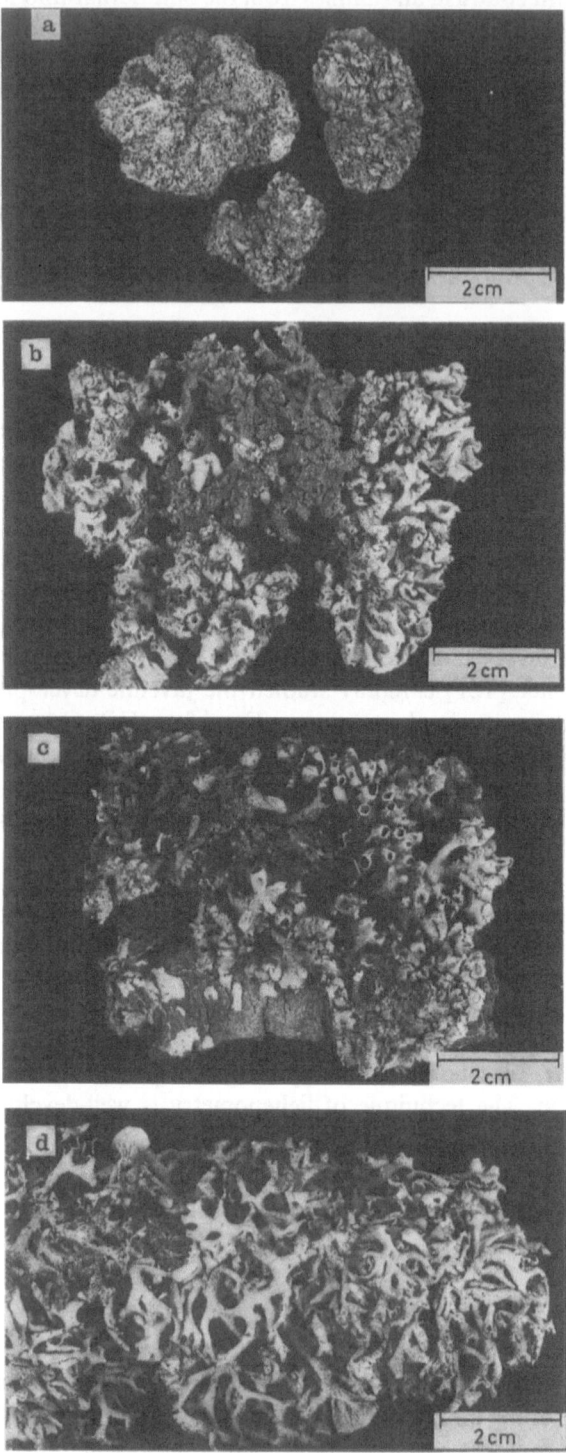

and soluble carbohydrates [108, 122], which can easily be monitored by measuring the conductivity of a leachate [110] or particularly the K^+ content of the leachate [111]. Brown and Slingsby [123] demonstrated that 65–85% of the total potassium in *Cladonia rangiformis* was intracellular and freely diffusible. Normally this K^+ is retained within the permeability barriers of the intact cell membrane and only a small portion is bound in an insoluble form or ionically to the cell wall. Potassium loss is described as a response to treatments with solutions of heavy metals and SO_2 and it varies in relation to pollutant concentration. The elements Hg, Ag, and Cu (class B and borderline elements with mostly class B characters—see Ref. [124]) induced a gradual response [125] with the highest K^+ loss induced by Hg (Fig. 7). A different response pattern was induced by Ni, Co, Cd, and Pb (class A and borderline with class A characters) treatments [125]. At low concentrations, K^+ loss increased slightly with increasing metal concentration until reaching a critical threshold above which further increases in metal concentration produced a sharp rise in K^+ loss (Fig. 7). The initial K^+ loss was interpreted as displacement of K^+ from ion-exchange sites [126, 127] and the subsequent abrupt increase to disruption of membrane integrity. Sulfur dioxide induces a third pattern of K^+ loss (Fig. 8), apparently related to two different SO_2 uptake systems.

Photosynthesis and Respiration

Effects of air pollutants on lichen net photosynthesis and respiration are measured as either CO_2 or O_2 gas exchange or in the case of gross photosynthesis as ^{14}C-fixation and recently with photoacoustic spectroscopy. Photosynthesis, of course, provides a response variable specific to the photobiont, whereas respiration is almost always dominated by the mycobiont. Differential sensitivity among species to different laboratory treatments is well demonstrated in this way, including studies with sulfite [128], SO_2 [32], ozone [63, 64], PAN [65], hydrogen fluoride [129], and metals [130]. In addition, these response variables were also used following field fumigation studies [131–133]. Comparative studies have found photosynthesis to be a more sensitive response variable than respiration [55] [134].

The effect of SO_2 on photosynthesis has been studied extensively [51, 135, 136]. At very low concentrations, a slight stimulation is reported, but higher concentrations cause a reduction (Fig. 9). Specific response is influenced by exposure time, pH [77], and hydration status [32]. Reduction of photosynthesis increased with exposure time (implying a dose-response relationship) and was more severe at lower pH and at higher thallus saturation. In cases where injury has not been severe, some photosynthetic recovery following cessation of treatments in both

---

◄ **Fig. 6.** Reference standards showing variation in thallus characteristics of *Hypogymnia enteromorpha* collected from sites in the Southern California mountains with varying oxidant levels: (**A**) bleached, convoluted, and reduced, San Bernardino Mountains; (**B**) moderately bleached, convoluted, and reduced, San Bernardino Mountains; (**C**) slightly bleached, convoluted, and reduced, San Bernardino Mountains; (**D**) unbleached and unconvoluted, Cuyamaca Rancho State Park. After [61]

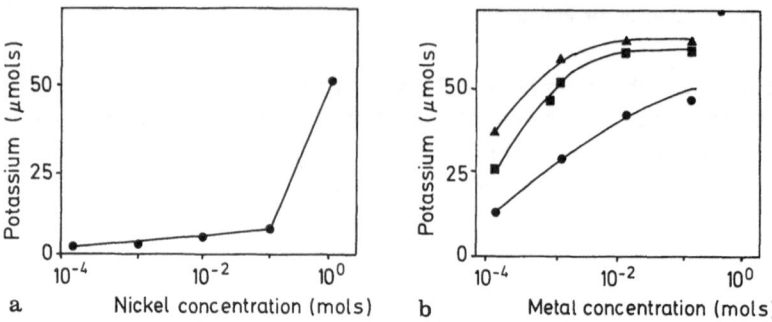

a   Nickel concentration (mols)   b   Metal concentration (mols)

Fig. 7. (**A**) The loss of potassium from *Umbilicaria mühlenbergii* as a function of the nickel concentration. Discs (120) of *U. mühlenbergii* were incubated for 3 h in solutions (50 ml) of nickel chloride which varied in concentration from 2×10^{-4} M to 2 M. The amount of potassium lost into the metal-ion solution was then determined. (**B**) The loss of potassium from *U. mühlenbergii* as a function of the copper, mercury, and silver concentration: ●, copper; ▲, mercury; ■, silver. Discs (120) of *U. mühlenbergii* were incubated in solutions of copper chloride, mercuric chloride, and silver nitrate which varied in concentration from 2×10^{-4} M to 2×10^{-1} M for 3 H. The amount of potassium lost into the medium was then determined. After [125]

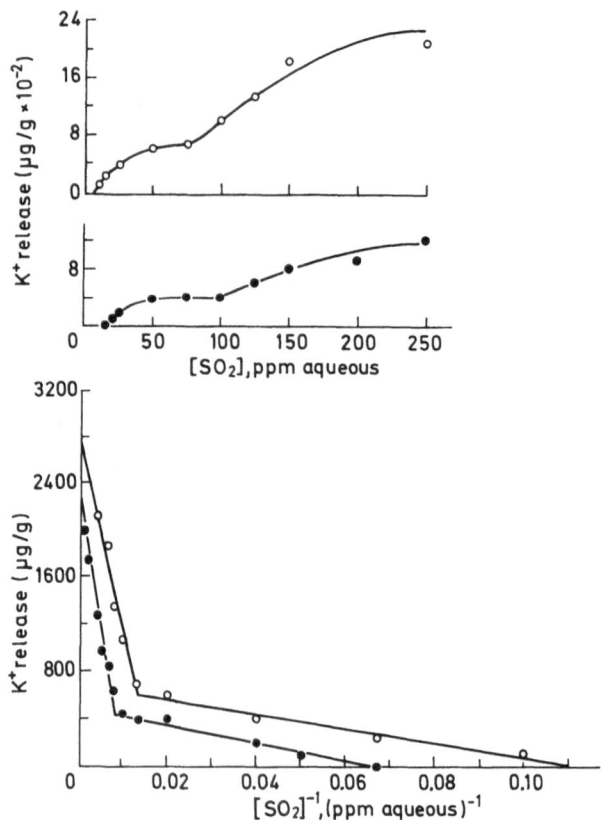

Fig. 8. Potassium release (μg/g oven-dry material) by *Cladonia rangiferina* and *Umbilicaria mühlenbergii* as a function of $[SO_2]^{-1}$ and $[SO_2]$ (inset); ▲, *Cladonia rangiferina*, ●, *Umbilicaria mühlenbergii*. Fifty 5-mm discs of *U. mühlenbergii* or 150 mg air-dry wt of *Cladonia rangiferina* were incubated for 3 h in 10 ml and 12 ml of aqueous SO_2 solution, respectively. Extrapolated aqueous threshold SO_2 concentrations were 15 ± 2 ppm (*U. mühlenbergii*) and 9 ± 1 ppm (*C. rangiferina*). After [51]

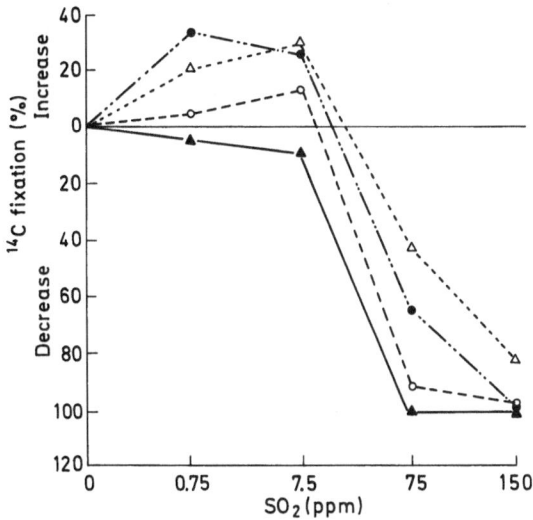

Fig. 9 The percentage reduction in net fixation by various lichens incubated in solutions of sulfur dioxide buffered at pH 4.4 as compared with control samples without sulfur dioxide. Each point represents the mean of six replicates. ■, *Cladonia alpestris*; ●, *C. deformis*; ○, *Umbilicaria mühlenbergii*; □, *Stereocaulon paschale*. After [51]

laboratory [137] and field [50] studies is demonstrated. Huebert et al. [138] recently emphasized the importance of thallus hydration. With a more advance method of keeping the thallus of *Evernia mesomorpha* at constantly high water contents, they detected reduction of net assimilation rate at their lowest SO_2 concentration (250 $\mu g\, m^3$). In contrast to earlier studies, they did not confirm a dependency of photosynthetic reduction with exposure time. Consequently, they suggested that peak concentrations of pollutants were more important than threshold concentrations for lichen survival.

Additional information on the light reactions of photosynthesis can be gained by the use of photoacoustic spectroscopy [139, 140]. Simultaneous with measuring gross photosynthesis, they obtained data on the individual photosystems by measuring oxygen evolution quantum yield and storage quantum yield. Differential reductions in the two processes due to SO_2 fumigation were interpreted as different mechanisms of injury to non-cyclic electron flow and in processes responsible for storing photochemical energy.

Different metals are also known to affect photosynthesis. Copper, silver, and mercury all caused a major reduction on ^{14}C fixation [125, 130, 141], while low concentrations (less than 12 $\mu mol\, g^{-1}$) of Sr, Mg, Ca, Ni, and Zn had no effect. In experiments with high Zn and Cd concentrations, differential sensitivity among species is well demonstrated. Nash [36] found net photosynthetic reductions at 0.1 mM Zn for *Cladonia uncialis* and *Lasallia papulosa*, while Brown and Beckett [130] reported a stimulation in *Dermatocarpon miniatum*, no effect in *Cladonia rangiformis*, and a substantial reduction in *Peltigera horizontalis*. The relative toxicity to photosynthesis was established by Puckett [125] as Ag, Hg, Co > Cu,

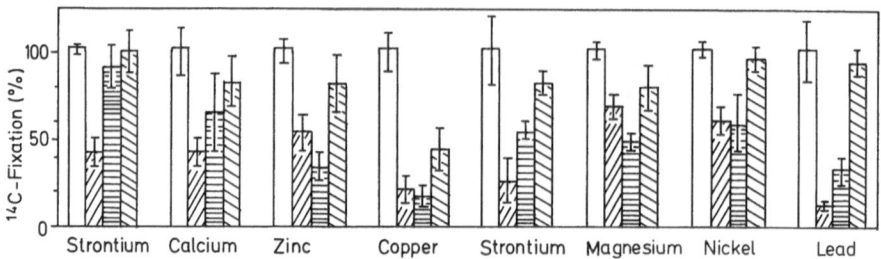

Fig. 10. Total ^{14}C fixation in samples of the lichen *Umbilicaria mühlenbergii* following metal uptake, SO$_2$ exposure, or the sequential administration of these. Histograms labelled SO$_2$ & M^{2+}, SO$_2$, M^{2+}, and control. Each set of histograms represents the data from experiments on separately collected lichen material. Error bars denote standard deviations (3 replicates). (a) The metal uptake levels and the total contents after the uptake step (in brackets) in μmol g^{-1} were: Sr, 5.3 (5.3); Ca, 11.5 (17.3); Zn, 16.5 (17.4); and Cu 12.3 (12.8). (b) The metal uptake levels and the total contents after the uptake step (in brackets) in μmol g^{-1} were: Sr, 10.4 (10.4); Mg, 11.3 (30.1); Ni, 10.4 (10.8); and Pb, 9.5 (9.7). □, Control; □, SO$_2$; □, M^{2+} and SO$_2$; □, M^{2+}. After [141]

Cd > Pb, Ni for short-term exposures, and Ag, Hg, Cu ≥ Pb, Co > Ni for extended exposure.

Combined SO$_2$ and metal treatments revealed unexpected results [141], as protection of photosynthesis against SO$_2$ injury by Sr and Ca was found. All other combinations showed a more-or-less cumulative effect on ^{14}C fixation (Fig. 10).

Differential photosynthetic sensitivity among lichen species is also reported for ozone fumigations. A one week fumigation period with up to 1567 μg m^3 had essentially no effect on *Cladonia stellaris* [66], while Ross and Nash [63] found a significant reduction in *Flavoparmelia caperata* at 200 μg m^3 for 12-h fumigations.

Pigment Degradation

Although a number of pigments are known from lichens, pigment degradation in air pollution studies has been restricted to the chlorophylls, the molecules responsible for light capture in photosynthesis. Thus, pigment degradation may be used as a surrogate variable for photosynthesis effects. Degradation may involve a bleaching of the thallus surface, a conversion of chlorophylls to their respective phaeophytins, a blue shift in the pigment absorption spectrum, and, in extreme cases, complete pigment elimination [51]. More recently, Kauppi [142] developed fluorescence microscopy techniques for investigating the initial events in pigment degradation. Under blue and UV excitation light, the photobiont usually exhibits a red primary fluorescence. However, after SO$_2$ fumigation, this fluorescence shifts initially to brown or orange and finally to white. In comparison to pigment extraction with acetone [30] or DMSO [143], the fluorescence procedure has the advantage of being more rapid and involving a smaller sample size and being non-destructive.

With vascular plants, a number of air pollutant-specific necrotic (brown patches of dead cells) and/or chlorotic (bleached cells) patterns are known. [4]; but with lichens, air pollutant-specific symptomology is not generally recognized, although general chlorotic and necrotic patterns are ascribed to SO_2 [30, 51], HF [33, 144, 145], oxidants [61], and Cu [125]. However, even in the absence of surface symptoms, some pigment degradation may occur [30, 143].

Nitrogen Fixation

Nitrogen fixation is, of course, restricted to lichen species containing cyano-bacteria. Although these species number less than 10% of all lichen species [10], cyanobacteria-containing lichens are abundant in a number of ecosystems and may account for a major fraction of the nitrogen fixed in these systems [18]. The well-known dramatic decline of the European Lobarion community [26, 40], that contains many large foliose nitrogen-fixing lichens, provides field evidence that air pollution effects on nitrogen fixation may be major.

Experimental evidence of the marked effect of acidity on nitrogen fixation is discussed in the section on "Acid Rain and Fog", and treatments with other air pollutants also have caused major effects on nitrogen fixation. Investigations with SO_2 [146] have demonstrated significant declines in nitrogen fixation at concentrations as low as 0.1 ppm SO_2. On the other hand, studies with nickel [147] and vanadium [148] documented stimulation of nitrogenase at low concentrations (12.5 ng Vl^{-1} and 50.0 mg Nil^{-1}), but at higher concentrations inhibition occurred. In other elemental studies, Henriksson and DaSilva [149] demonstrated that low concentrations ($5–25$ ngl^{-1}) of As, Cd, Pb, Ni, Pd, and Zn stimulated nitrogen fixation in the isolated photobiont of the common soil lichen *Collema tenax*, but that higher concentrations ($25–125$ ngl^{-1}) were inhibitory. Furthermore, Sheridan [150] designed experiments to mimic conditions around a coal-fired power plant. He documented significant reduction in nitrogen fixation by mixtures of fluoride (10 ngg^{-1}) and lead ($5.3:1$ ratio) and bisulphite (0.1 $\mu g\,g^{-1}$).

Other Metabolic Parameters

A variety of physiological and biochemical factors have been measured as response variables in lichens to air pollutant exposure. For example, the activity of phosphatase, an enzyme whose extracellular localization [151] makes it very susceptible to changes in its immediate environment, has been studied in relation to several pollutants. For the lichen *Cladonia rangiferina*, Lane and Puckett [152] showed reduction in acitivity caused by uranyl, vanadyl, biselenite, cyanide, fluoride, molybdate, phosphate, and vanadate anions as well as stimulation by copper, nickel, and silver. Reductions in activity were also demonstrated for field studies in Germany [153] and following gaseous fumigations [154].

The energy status of cells as documented by adenine nucleotide status also is influenced by air pollutants. With SO_2 exposures, Köck and Schlee [155] documented a shift from ATP to AMP. Kardish et al. [156] found that this shift was detectable prior to chlorophyll degradation. Protein and lipid biosynthesis is also affected by SO_2. Malhotra and Khan [157] measured reduction in both processes down to 0.1 ppm SO_2 and recovery following cessation of fumigation only occurred after 0.1 ppm SO_2 was employed. In contrast to plants under drought and salt stress, reduction in proline production was shown in the lichen alga *Trebouxia* under SO_2 stress [158]. These observations may result from sulfite inhibition of proline dehydrogenase.

Ultrastructure

Green algae-containing lichens

Changes in ultrastructure, as documented by electron microscope studies, are demonstrated both for lichens collected at polluted sites in the field [112, 159] and after controlled fumigations with SO_2 and O_3 [58] and PAN [160]. Early stages of injury were characterized by changes of chloroplast and photobiont's mitochondrial structures. Eversman and Sigal [58, 160] also found increased accumulation of starch and cavity space around starch grains in the chloroplast. Holopainen and Kärenlampi [161] described a stretching of the chloroplast envelope and a curled appearance of the thylakoids, with starch accumulation at higher SO_2 concentrations. The algal mitochondria were swollen and deformed at this stage. With increased pollution stress, deformation of pyrenoglobuli and decreased size of pyrenoids occurred. In the final stages, no cell organelles were discernible; only starch grains and cytoplasmic storage droplets remained.

In contrast, the fungal cell structures were clearly more pollution-tolerant. Their mitochondria showed the same symptoms as the algal mitochondria, but SO_2 concentrations almost an order of magnitude higher were required to induce these effects [161]. Other symptoms of injury in the fungal cells were small storage droplets, many vacuoles, and dark inclusions [159]. At low levels of pollution stress, ultrastructural change was more species-specific than gaseous pollutant-specific [58]. The concentration at which ultrastructural changes occurred was species-specific and was also dependent on water status [161] and light conditions during fumigation [58].

Cyanobacteria-containing lichens

In the cyanobacterium *Nostoc* from *Peltigera canina*, increasing SO_2 concentrations resulted in marked decrease in the number of carboxysomes [162]. At higher SO_2 levels, increases in the size and number of cyanophycin granules, the nitrogen reserves, were found. Polyglucoside granules became apparent among the thylakoid membranes during akinete differentiation, which itself is unusual for *Nostoc* living in symbiosis.

Concluding Remarks

Clearly, lichens are good indicators of air pollution, particularly for sulfur dioxide. The supporting literature is extensive and new updates are provided several times a year in the British journal, The Lichenologist. Our knowledge is, however, uneven. Oxidants, the pollutants assumed to be most important to vascular plants, have received comparatively little attention compared to SO_2 in the lichenological literature. The discrepancy of probable effects occurring in the field, but poor corroboration in laboratory studies should be one area of focus in the future.

In addition, there are a range of contaminants for which we have essentially no information on lichen sensitivities. Perhaps the most prominent gap today lies in the broad classes of air pollutants collectively referred to as "the organics". For example, there is some evidence of long range transport of organics from temperate latitudes to polar regions where they recondense and become deposited on lichens [163]. The effect, if any, of these compounds is unknown, although oil spill effects have been investigated [164, 165]. Lichens have been used to study deposition patterns of chlorinated hydrocarbons [163, 166, 167], polychlorinated biphenyls [168], and PAH compounds and phthalates [163].

Finally, there is a need for governmental regulatory agencies to become more involved with lichenological studies. The relative lack of such interaction between government and scientists has led to the unfortunate fact that the lichenological literature has not been used by such agencies as extensively as it deserves [169]. Obviously there is a need not only to fund additional projects, but also to utilize peer review of designs for such projects so that maximal beneficial information is obtained. A much more extensive database on dose-response relationships is one prominent example of additional work needed by governmental agencies.

References

1. Nash III TH, Wirth V (eds) (1988) Lichens, Bryophytes and Air Quality, Bibliotheca Licheno-logica, Vol 30. J. Cramer, Berlin, Stuttgart
2. Richardson DHS (1988) Bot J Linnean Soc 96: 31
3. Hawksworth DL, Rose F (1976) Lichens as Pollution Monitors, Studies in Biology no 66. E. Arnold, London
4. Jacobson JS, Hill AC (1970) Recognition of Air Pollution Injury to Vegetation: A Pictorial Atlas. Air Pollution Control Association, Pittsburg, p A1-H3
5. Puckett KJ (1988) Bryophytes and lichens as monitors of metal deposition, in: Nash III TH, Wirth V (eds) Lichens, Bryophytes and Air Quality. J. Cramer, Berlin, Stuttgart, p 231
6. Boonpragob K, Nash III TH, Fox CA (1989) Expt Environ Bot 29: 187
7. Boonpragob K, Nash III TH (1990) Expt Environ Bot 30: in press
8. Haagen-Smit AJ (1952) Ind Eng Chem 44: 1342
9. Hale Jr ME (1983) The Biology of Lichens, E. Arnold, London
10. Hawksworth DL, Hill DJ (1984) The Lichen-forming Fungi, Blackie, Glasgow, London
11. Henssen A, Jahns HM (1974) Lichenes. Eine Einführung in die Flechtenkunde, G. Thieme, Stuttgart
12. Galun M (ed) (1988) Handbook of Lichenology, Vol. I–III, CRC Press, Boca Raton
13. Ahmadjian V, Hale Jr ME (eds) (1973) The Lichens, Academic Press, New York
14. Brown DH, Hawksworth DL, Bailey RH (eds) (1976) Lichenology: Progress and Problems, Academic Press, London
15. Brown DH (ed) (1985) Lichen Physiology and Cell Biology, Plenum, New York, London
16. Lawrey JD (1984) Biology of Lichenized Fungi, Praeger, New York

17. Ahmadjian V, Jacobs JB (1983) Algal-fungal relationships in lichens: recognition, synthesis, and development, in: Goff L (ed) Algal Symbioses: A Continuum of Interaction Strategies, Cambridge, New York, p 147
18. Slack NG (1988) The ecological importance of lichens and bryophytes, in: Nash III TH, Wirth V (eds) Lichens, Bryophytes and Air Quality. J. Cramer, Berlin, Stuttgart, p 23
19. Seaward MRD (1988) Contribution of Lichens to Ecosystems, in: Galun M (ed) Handbook of Lichenology II. CRC Press, Boca Raton, p 107
20. Larson DW (1987) The absorption and release of water by lichens, in: Peveling E (ed) Progress and Problems in Lichenology in the Eighties, Bibliotheca Lichenologica 25. J. Cramer, Berlin, Stuttgart, p 351
21. Pike LH (1978) Bryologist 81: 247
22. Hawksworth DL (1971) Int J Environ Studies 1: 281
23. Sernander R (1912) Svensk Bot Tidskr 6: 803
24. Sernander R (1926) Stockholms Natur, Almqvist and Wiksells, Uppsala
25. Showman RE (1988) Mapping air quality with lichens, the North American experience, in: Nash III TH, Wirth V (eds) Lichens, Bryophytes and Air Quality. J. Cramer, Berlin, Stuttgart, p 67
26. Wirth V (1988) Phytosociological approaches to air pollution monitoring with lichens, in: Nash III TH, Wirth V (eds) Lichens, Bryophytes and Air Quality. J. Cramer, Berlin, Stuttgart, p 91
27. Will-Wolf S (1988) Quantitative approaches to air quality studies, in: Nash III TH, Wirth V (eds) Lichens, Bryophytes and Air Quality. J. Cramer, Berlin, Stuttgart, p 109
28. Nash III TH (1976) Naturwissenschaften 63: 364
29. Ferry BW, Baddeley MS, Hawksworth DL (1973) Air Pollution and Lichens. Athlone, London
30. Nash III TH (1973) Bryologist 76: 333
31. Baddeley MS, Ferry BW, Finegan EJ (1973) Sulphur dioxide and respiration in lichens, in: Ferry BW, Baddeley MS, Hawksworth DL (eds) Air Pollution and Lichens. Athlone, London, p 299
32. Türk R, Wirth V, Lange OL (1974) Oecologia 15: 33
33. Nash III TH (1971) Bull Torrey Bot Club 98: 103
34. Gilbert OL (1971) Lichenologist 5: 26
35. Gilbert OL (1973) The Effect of Airborne Fluorides, in: Ferry BW, Baddeley MS, Hawksworth DL (eds) Air Pollution and Lichens. Athlone, London, p 176
36. Nash III TH (1975) Ecol Mongr 45: 183
37. Hawksworth DL, Rose F, Coppins BJ (1973) Changes in the lichen flora of England and Wales attributable to pollution of the air by sulphur dioxide, in: Ferry BW, Baddeley MS, Hawksworth DL (eds) Air Pollution and Lichens. Athlone, London, p 330
38. de Wit T (1976) Epiphytic Lichens and Air Pollution in The Netherlands, Bibliotheca Lichenologica 5. J. Cramer, Berlin, Stuttgart
39. Wirth V (1976) Schriftenreihe Vegetationsk 10: 177
40. Hallingbach T (1986) Svensk Bot Tidsk 80: 373
41. Hawksworth DL, Rose F (1970) Nature 227: 145
42. Gilbert OL (1970) New Phytol 69: 605
43. Nash III TH (1988) Correlating fumigation studies with field effects, in: Nash III TH, Wirth V (eds) Lichens, Bryophytes and Air Quality. J. Cramer, Berlin, Stuttgart, p 201
44. Rose CI, Hawksworth DL (1981) Nature 289: 289
45. Henderson-Sellers A, Seaward MRD (1979) Environ Pollut 19: 207
46. Seaward MRD (1982) Urban Ecology 7: 181
47. Muir PS, McCune B (1988) J Environ Qual 17: 361
48. Winner WE, Atkinson CJ, Nash III TH (1988) Comparisons of SO_2 absorption capacities of mosses, lichens, and vascular plants in diverse habitats, in: Nash III TH, Wirth V (eds) Lichens, Bryophytes and Air Quality. J. Cramer, Berlin, Stuttgart, p 217
49. Grace B, Gillespie TJ, Puckett KJ (1985) Canad J Bot 63: 797
50. Moser TJ, Nash III TH, Link SO (1983) Canad J Bot 61: 642
51. Puckett KJ, Nieboer E, Flora WP, Richardson DHS (1973) New Phytol 72: 141
52. Richardson DHS, Nieboer E (1983) J Hattori Gard 54: 331
53. Fields RF (1988) Physiological responses of lichens to air pollutant fumigations, in: Nash III TH, Wirth V (eds) Lichens, Bryophytes and Air Quality. J. Cramer, Berlin, Stuttgart, p 175

54. Ziegler I (1977) Oecologia 29: 63
55. Köck M, Schlee D, Metzger U (1985) Biochem Physiol Pflanzen 180: 213
56. Marti J (1985) Die Toxizität von Zink, Schwefel- und Stickstoff-verbindungen auf Flechten-Symbionten, Bibliotheca Lichenologica *21*. J. Cramer, Vaduz
57. Fields RF, St Clair LL (1984) Bryologist 97: 297
58. Eversman S, Sigal LL (1987) Canad J Bot 65: 1806
59. National Research Council (1977) Ozone and other photochemical oxidants. National Academy of Sciences, Washington D.C.
60. Ross LJ (1982) Lichens on Coastal Live Oak in Relation to Ozone. MS thesis, Arizona State University
61. Sigal LL, Nash III TH (1983) Ecology 64: 1343
62. Nash III TH, Sigal LL (1980) Sensitivity of lichens to air pollution with an emphasis on oxidant air pollutants, in: Miller PR (ed) Proceedings of the Symposium on Effects of Air Pollutants on Mediterranean and Temperate Forest Ecosystems. U.S. Department of Agriculture, Berkeley, p 117
63. Ross LJ, Nash III TH (1983) Environ Expt Bot 23: 71
64. Nash III TH, Sigal LL (1979) Bryologist 82: 280
65. Sigal LL, Taylor OC (1979) Bryologist 82: 564
66. Rosentreter R, Ahmadjian V (1977) Bryologist 80: 600
67. Brown DH, Smirnoff N (1978) Lichenologist 10: 91
68. McCune B (1988) Bryologist 91: 223
69. Waldman JM (1984) J Air Pollut Contr Ass 34: 13
70. Skye E (1968) Acta Phytogeogr Suecica 52: 1
71. Johnsen I, Søchting U (1976) Bryologist 79: 86
72. Kiss T (1987) Symposia Biol Hungarica 35: 865
73. Robitaille G, LeBlanc F, Rao DN (1977) Rev Bryol Lichénol 43: 53
74. Gilbert OL (1986) Environ Pollut 40: 227
75. LeBlanc F, Rabitaille G, Rao DN (1974) J Hattori Bot Lab 38: 405
76. Hällgren JE, Huss K (1975) Physiol Plant 34: 171
77. Türk R, Wirth V (1975) Oecologia 19: 285
78. Sigal LL, Johnston Jr JW (1986) Water Air Soil Pollut 27: 315
79. Bailey C, Larson DW (1982) Bryologist 85: 431
80. Lechowicz MJ (1982) Water Air Soil Pollut 18: 421
81. Lechowicz MJ (1987) Water Air Soil Pollut 34: 71
82. Sigal LL, Johnston Jr JW (1986) Environ Expt Bot 26: 59
83. Denison R, Caldwell B, Bormann B, Eldred L, Swanberg C, Anderson S (1977) Water Air Soil Pollut 8: 21
84. Gunther AJ (1988) Water Air Soil Pollut 38: 379
85. Garty J, Galun M, Kessel M (1979) New Phytol 82: 159
86. Nieboer E, Richardson DHS (1981) Lichens as monitors of atmospheric deposition, in: Eisenreich SJ (ed) Atmospheric Pollutants in Natural Waters. Ann Arbor Science, Ann Arbor, p 339
87. Collins CR, Farrar JF (1978) New Phytol 81: 71
88. Brown DH, Beckett RP (1985) The role of the cell wall in the intracellular uptake of cations by lichens, in: Brown DH (ed) Lichen Physiology and Cell Biology, Plenum, New York, London, p 247
89. Brown DH, Beckett RP (1984) Lichenologist 16: 173
90. Richardson DHS, Nieboer E (1980) Surface binding and accumulation of metals in lichens, in: Cook CB, Pappas PW, Rudolph ED (eds) Cellular Interactions in Symbiosis and Parasitism. Ohio State University, Columbus, p 75
91. Nieboer E, Richardson DHS, Tomassini FD (1978) Bryologist 81: 226
92. Jones D, Wilson MJ, Laundon JR (1982) Lichenologist 14: 281
93. Nash III TH (1990) Tolerance of lichens to heavy metals, in: Shaw J (ed) Heavy Metal Tolerance in Plants: Evolutionary Aspects. CRC Press, Boca Raton, in press
94. Lange OL, Ziegler H (1963) Mitt flor-soz Arbeitsgem NF 10: 156
95. Purvis OW (1984) Lichenologist 16: 197

96. Purvis OW, Elix JA, Broomhead JA, Jones GC (1987) Lichenologist 19: 193
97. Nash III TH (1989) Tolerance of lichens to heavy metals, in: Shaw J (ed) Heavy Metal Tolerance in Plants: Evolutionary Aspects. CRC Press, Boca Raton, in press
98. Puckett KJ, Finegan EJ (1981) Source identification of the various elements accumulated by lichens in the vicinity of St. John, New Brunswick. in: Proc. Int. Conf. Heavy Metals in the Environment, CEP Consultants, Edinburgh, p 326
99. Lang GE, Reiners WA, Heier RK (1976) Oecologia 25: 229
100. Olson RK, Reiners WA, Cronan CS, Lang GE (1981) Holarc Ecol 4: 291
101. Reiners WA, Olson RK (1984) Oecologia 63: 320
102. Boucher VL, Nash III TH (1990) Bot Gaz, in press
103. Lovett GM, Lindberg, SE (1984) J Appl Ecol 21: 1013
104. Tukey HB (1970) Ann Rev Plant Physiol 21: 305
105. Boucher VL, Nash III TH (1989) Bot Gaz, in press
106. Parker GG (1983) Adv Ecol Res 13: 57
107. Boonpragob K, Nash III TH (1990) Environ Expt Bot, in press
108. Brown DH, Buck GW (1979) New Phytol 82: 115
109. Puckett KJ, Burton MAS (1981) The effect of trace elements on lower plants, in: Lepp NW (ed) Effect of Heavy Metal Pollution on Plants. Vol 2. Applied Sci, London, p 213
110. Pearson LC, Henriksson E (1981) Bryologist 84: 515
111. Puckett KJ, Tomassini FD, Nieboer E, Richardson DHS (1977) New Phytol 79: 135
112. Holopainen T (1982) Saronia 5: 15
113. LeBlanc F, De Sloover J (1970) Canad J Bot 48: 1485
114. Pyatt FB (1973) Lichen propagules, in: Ahmadjian V, Hale ME (eds) The Lichens, Academic Press, New York, London, p 117
115. Kofler L, Villiot ML, Fontanges R (1972) Sciences III: 170
116. Belandria G, Asta J, Nurit F (1989) Lichenologist 21: 79
117. LeBlanc F, Rao DN (1973) Ecology 54: 612
118. Margot J (1973) Experimental study of the effects of sulphur dioxide on the soredia of *Hypogymnia physodes*, in: Ferry BW, Baddeley MS, Hawksworth DL (eds) Air Pollution and Lichens, Athlone Press, London, p 314
119. Goyal R, Seaward MRD (1982) New Phytol 90: 73
120. Lawrey JD, Hale Jr ME (1979) Science 204: 423
121. Schuster G (1985) Die Jugendentwicklung von Flechten, Bibliotheca Lichenologica 20. J. Cramer, Vaduz
122. Dudley SA, Lechowicz MJ (1987) Plant Physiol 83: 813
123. Brown DH, Slingsby DR (1972) New Phytol 71: 297
124. Nieboer E, Richardson DHS (1980) Environ Poll B 1: 3
125. Puckett KJ (1976) Canad J Bot 54: 2695
126. Nieboer E, Lavoie P, Sasseville RLP, Puckett KJ, Richardson DHS (1976) Canad J Bot 54: 720
127. Nieboer E, Puckett KJ, Grace B (1976) Canad J Bot 54: 724
128. Hill DJ (1974) New Phytologist 73: 1193
129. Börtitz S, Ranft H (1972) Biol. Zentralblatt 91: 613
130. Brown DH, Beckett RP (1983) Ann Bot 52: 51
131. Moser TJ, Nash III TH, Clark WD (1980) Canad J Bot 58: 2235
132. Moser TJ, Nash III, Olafsen AG (1983) Canad J Bot 61: 367
133. Eversman S (1978) Bryologist 81: 368
134. Beekley PK, Hoffman GR (1981) Bryologist 84: 379
135. Puckett KJ, Richardson DHS, Flora WP, Nieboer E (1974) New Phytol 73: 1183
136. Hill DJ (1971) New Phytol 70: 831
137. Coxson DS (1988) New Phytol 108: 483
138. Huebert DB, L'Hirondelle SJ, Addison PA (1985) New Phytol 100: 643
139. Ronen R, Canaani O, Garty J, Cahen D, Malkin S, Galun M (1984) Adv Photosyn Res 4: 1
140. Ronen R, Canaani O, Garty J, Cahen D, Malkin S, Galun M (1985) Photosynthetic parameters in *Ramalina duriaei*, in vivo, studied by photoacoustics, in: Brown DH (ed) Lichen Physiology and Cell Biology, Plenum, New York, p 9

141. Richardson DHS, Nieboer E, Lavoie P, Padovan D (1979) New Phytol 82: 633
142. Kauppi M (1980) Ann Bot Fennici 17: 163
143. Ronen R, Galun M (1984) Environ Expt Bot 24: 239
144. LeBlanc F, Rao DN, Comeau G (1972) Canad J Bot 50: 991
145. Perkins DF, Millar RO (1987) Environ Pollut 47: 63
146. Henriksson E, Pearson LC (1981) Amer J Bot 68: 680
147. Rai LC, Raizada M (1986) New Phytol 104: 111
148. Vaishampayan A (1983) New Phytol 95: 55
149. Henriksson LE, DaSilva EJ (1978) Zeitschr Allg Mikrobiol 18: 487
150. Sheridan RP (1979) Bryologist 82: 54
151. Boissière MC (1973) CR Acad Sci Paris 277: 1649
152. Lane I, Puckett KJ (1979) Canad J Bot 57: 1534
153. Bauer E, Kreeb K (1974) Verh Ges Ökologie, Saarbrücken 1973: 273
154. Schmid ML, Kreeb K (1975) Angew Bot 49: 141
155. Köck M, Schlee D (1981) Phytochem 20: 2089
156. Kardish N, Ronen R, Bubrick P, Garty J (1987) New Phytol 106: 697
157. Malhotra SS, Kahn AA (1983) Biochem Physiol Pflanzen 178: 121
158. Ewald D, Schlee D (1983) New Phytol 94: 235
159. Holopainen T (1981) Silva Fennica 15: 469
160. Eversman S, Sigal LL (1984) Bryologist 87: 112
161. Holopainen T, Kärenlampi L (1984) New Phytol 98: 285
162. Sharma P, Bergman B, Hällbom L, Hofsten AV (1982) New Phytol 92: 573
163. Carlberg GE, Ofstad EB, Drangsholt H, Steinnes E (1983) Chemosphere 12: 341
164. Brown DH (1972) Marine Biol 12: 309
165. Brown DH (1973) Marine Biol 18: 291
166. Villeneuve JP, Holm E (1984) Chemosphere 13: 1133
167. Villeneuve JP, Fogelqvist E, Cattini C (1988) Chemosphere 17: 399
168. Garty J, Perry AS, Mozel J (1982) Nord J Bot 2: 583
169. Sigal LL (1988) The relationship of lichen and bryophyte research to regulatory decisions in the United States. In: Nash III TH, Wirth V (eds) Lichens, Bryophytes and Air Quality. J. Cramer, Berlin, Stuttgart, p 269

Morbidity Associated with Air Pollution

Morton Lippmann

Institute of Environmental Medicine, New York University Medical Center, Tuxedo, NY 10987, USA

Summary

Numerous population studies have shown associations between community air pollution levels and various indices of morbidity. Mixtures of sulfur oxides and particulates (sulfur dioxide vapors, sulfuric acid and its ammonium salts, and other particulate matter) have been associated with increased rates of respiratory symptoms, hospital admissions for respiratory diseases, clinic visits, work-days lost, and persistent reductions in lung function. Ozone has been associated with transient reductions in lung function in healthy persons and increased symptom rates among persons with asthma. Carbon monoxide causes more rapid onset of angina and electrocardiographic pattern abnormalities among cardiovascular patients. Lead has been associated with a variety of neurobehavioral and functional deficits in children, including reductions in intelligence quotient, delayed developmental rates, increased postural sway, and reduced hearing acuity. In adults, lead has been associated with elevated blood pressures.

This chapter reviews the evidence for these associations and the extent of the population that may be affected. It excludes the effects of air pollutants on carcinogenesis on the basis that most of these responses are more appropriate to a chapter on mortality.

Ozone

Introduction

Ozone (O_3) exposure affects the structure and functions of the respiratory tract in a variety of ways. Most O_3 research on humans has focused on effects on respiratory function, especially on transient responses to acute exposures. Other lung functional responses to acute and subacute exposures that have been studied, largely in animals, include mucociliary and early alveolar zone particle clearance, functional responses in macrophages and epithelial cells, and changes in lung cell secretions. Structural changes in the smaller conductive airways and the more proximal gas exchange region have been associated with subchronic and chronic animal exposure protocols. However, the health significance of localized structural changes, and the roles of the transient functional and cellular responses, if any, in the pathogenesis of lung disease, remain speculative. The responses in animals suggest possible links between O_3 exposure and respiratory disease in humans, but do not provide clear evidence for such ties. The following summarizes current knowledge on O_3 formation and transport, human exposures, and transient and chronic health effects produced by the inhalation of O_3.

Formation and Transport Within the Atmosphere

Ozone (O_3) is a product of a series of photochemical reaction sequences that require hydrocarbon vapors (HC), nitrogen dioxide (NO_2), and sunlight. The O_3 concentrations usually peak in the late morning or afternoon and decline in the evening due to O_3 loss from reaction with nitric oxide (NO) and terrestrial surfaces. Since HC and NO_2 remain in the atmosphere, O_3 continues to form in the atmosphere far downwind of the sources. The O_3 concentrations are often higher in the suburbs and rural downwind areas than in the cities [1].

Population Exposures to Ozone

Widespread O_3 contamination increases both the number of people exposed and the magnitude of their exposure, especially in mild weather, when the largest number of people are outdoors for extended periods. Approximately half of the U.S. population lived in communities with ambient ozone concentrations that exceeded 240 $\mu g\,m^{-3}$ during the period 1983–1985 [2, 3]. More than 10 million people in the U.S. were estimated to be exposed to such concentrations while exercising at moderate to heavy levels of exertion [3]. In 1987 and 1988, ozone concentrations were higher than those in the 1983–1985 period, and the population overexposed was therefore larger.

The current U.S. National Ambient Air Quality Standard (NAAQS) of 240 $\mu g\,m^{-3}$ for O_3 uses a 1 h averaging time based primarily on the expectation that ambient exposures are characterized by relatively sharp afternoon peaks. However, it has recently been shown that ambient O_3 concentrations in the Netherlands and New Jersey often have broad daytime peaks, with maximum 8 h averages close to 90% of peak 1 h levels [4]. In the ambient air in the U.S. as a whole, a 1 h O_3 peak at 240 $\mu g\,m^{-3}$ is, on average, associated with a maximum 8 h average concentration of 200 $\mu g\,m^{-3}$ [2].

Transient Effects on Respiratory Function

It is well established that the inhalation of O_3 causes concentration dependent mean decrements in exhaled volumes and flow rates during forced expiratory maneuvers, and that the mean decrements increase with increasing depth of breathing [5]. There is a wide range of reproducible responsiveness among healthy subjects [6], and functional responsiveness to O_3 is no greater, and usually lower, among cigarette smokers [7, 8], older adults [9, 10], asthmatics [11, 12], and patients with allergic rhinitis [13] or chronic obstructive pulmonary disease (COPD) [14, 15]. It is also well established that repetitive daily exposures, at a level that produces a functional response upon single exposure, result in an enhanced response on the second day, with diminishing responses on days 3 and 4, and virtually no response by day 5 [16, 17, 18]. This functional adaptation to exposure disappears about a week after exposure ceases [19, 20]. The adaptation phenomenon has led some people to conclude that transient functional decrements are not important health effects. On the other hand, recent research in animals has shown that persistent damage to lung cells accumulates even as functional adaptation takes place. Tepper et al. [21] exposed rats to 700, 1,000, or 2,000 $\mu g\,m^{-3}$ O_3 for 2.25 h on five consecutive days. Carbon dioxide (8%) was added to the exposure during alternate 15 min periods to stimulate breathing and thereby increase O_3 uptake and distribution. The consequences of exposure on pulmonary function, histology, macrophage phagocytosis, lavagable protein, differential cell counts and lung tissue antioxidants were assessed. Tidal volume, frequency of breathing, inspiratory time, expiratory time and maximal tidal flows were affected by O_3 during day 1 and 2 at all O_3 concentrations. By day 5, these O_3 responses were completely adapted at 700 $\mu g\,m^{-3}$, greatly attenuated at 1,000 $\mu g\,m^{-3}$, but showed no signs of adaptation in the group exposed to 2,000 $\mu g\,m^{-3}$. Unlike the pulmonary function data, light microscopy indicated a pattern of progressive epithelial damage and inflammatory changes associated with the terminal bronchiole region. Over the five day testing period, a sustained 37% increase in lavagable protein and 60% suppression of macrophage phagocytic activity was observed with exposure to 1,000 $\mu g\,m^{-3}$. There were no changes in differential cell counts. Lung glutathione was initially increased, but was within the control range on days 4 and 5. Lung ascorbate was significantly elevated above control on days 3–5. These data suggest that attenuation of the pulmonary functional response occurs while aspects of the tissue response reveal progressive damage.

The first indications that the effects of O_3 on respiratory function accumulate over more than 1 h were the observations of McDonnell et al. [22] and Kulle et al. [23] in chamber exposures to O_3 in purified air for 2 h with the volunteer subjects engaged in vigorous intermittent exercise. Significant function decrements observed after 2 h of exposure were not present at measurements made after 1 h.

Spektor et al. noted that children at summer camps with active outdoor recreation programs had greater decrements in lung function than children exposed to O_3 at comparable concentrations in chambers for 1 or 2 h [24]. Furthermore, their activity levels, although not measured, were known to be considerably lower than those of the children exposed in the chamber studies while performing very vigorous exercise. Since it is well established that functional responses to O_3 increase with levels of physical activity and ventilation [5], the greater responses in the camp children had to be caused by other factors, such as greater cumulative exposure, or to the potentiation of the response to O_3 by other pollutants in the ambient air. Cumulative daily exposures to O_3 were generally greater for the camp children, since they were exposed all day long rather than for a 1 or 2 h period preceded and followed by clean air exposure.

Similar considerations apply to the study of Kinney et al. [25] of school children in Kingston and Harriman, TN, whose lung function was measured in school on up to six occasions during a 2 month period in the late winter and early spring. Child specific regressions of function versus maximum 1 h O_3 during the previous day indicated significant associations between O_3 and function, with coefficients similar to those seen in the summer camp studies of Lippmann et al. [26] and Spektor et al. [24]. Since children in school may be expected to have relatively low activity levels, the relatively high response coefficients may be due to potentiation by other pollutants or to a low-level of seasonal adaptation. Kingston-Harriman is notable for its relatively high levels of aerosol acidity. As shown by Spengler et al. [27], Kingston-Harriman has higher annual average and higher peak acid aerosol concentrations than other cities studied, i.e. Steubenville, OH; St. Louis , MO; and Portage, WI. Alternatively, the relatively high response coefficients could have been due to the fact that the measurements were made in the late winter and early spring. Linn et al. [28] have shown evidence for a seasonal adaptation, and children studied during the summer may not be as responsive as children measured earlier in the year.

In the study by Linn et al. [28] in Southern California, a group of subjects selected for their relatively high functional responsiveness to O_3 had much greater functional decrements following 2 h of exposure to O_3 at 360 μg m^{-3} with intermittent exercise in a chamber in the spring than they did in the following autumn or winter, while their responses in the following spring were equivalent to those in the preceding spring. These findings suggest that some of the variability in response coefficients reported for earlier controlled human exposures to O_3 in chambers could have been due to seasonal variations in responsiveness which, in turn, may be related to a long-term adaptation to chronic O_3 exposure.

The observations from the field studies in the children's camps stimulated Folinsbee et al. [29] to undertake a chamber exposure study of adult volunteers involving 6.6 h of O_3 exposure @ 240 μg m^{-3}. Moderate exercise was performed for 50 min h^{-1} for 3 h in the morning, and again in the afternoon. They found that the function decrements become progressively greater after each hour of exposure, reaching average values of ~ 400 mL for forced vital capacity (FVC) and ~ 540 mL for forced expiratory volume in one second (FEV$_1$) by the end of the day (Fig. 1). The effects were transient in that there were no residual function decrements on the following day. The decrements in FEV$_1$ after 6.6 h of exposure at

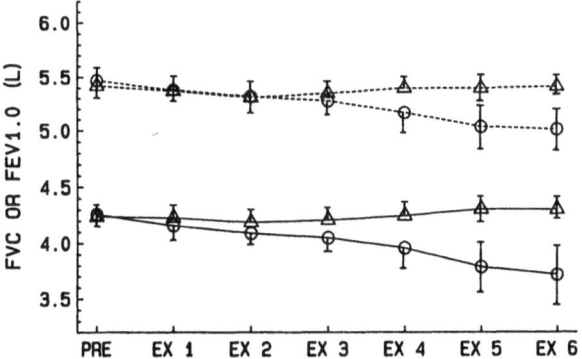

Fig. 1. Mean $\pm$ S.E. of forced vital capacity and forced expired volume in one second after each of six 1 h exercise periods. *Dashed lines* are FVC and *solid lines* are FEV$_1$; *triangles* represent air exposures and *circles* ozone exposures. From: Folinsbee et al. [29]

$240\,\mu\mathrm{g\,m}^{-3}$ averaged 13%, and were comparable to those seen previously in the same laboratory on similar subjects following 2 h of intermittent heavier exercise (68 liters inhaled per minute for a total exercise time of 60 min) at an interpolated concentration of $\sim 430\,\mu\mathrm{g\,m}^{-3}$. Assuming that the rate of ventilation was 10 liters per min between exercise periods, the total amount of O_3 inhaled during 2 h of intermittent heavy exercise at $430\,\mu\mathrm{g\,m}^{-3}$ would be $[60\,\min \times 0.068\,\mathrm{m}^3\,\min^{-1} + 60\,\min \times 0.010\,\mathrm{m}^3\,\min^{-1} \times 430\,\mu\mathrm{g\,m}^{-3} = 2.01\,\mathrm{mg}$ O_3. The corresponding amount of O_3 inhaled during 6.6 h of intermittent moderate exercise at $235\,\mu\mathrm{g\,m}^{-3}$ would be $[300\,\min \times 0.040\,\mathrm{mg\,m}^{-3} + 100\,\min \times 0.010\,\mathrm{m}^3\,\min^{-1} \times 235\,\mu\mathrm{g\,m}^{-3}$ $= 3.06\,\mathrm{mg}$ O_3. Thus, the effect accumulates with time, but there appears to be a temporal decay of effect going on at the same time. Follow-up studies in the same laboratory were done with 6.6 h exposures at 160, 200, and $240\,\mu\mathrm{g\,m}^{-3}$. The results at $240\,\mu\mathrm{g\,m}^{-3}$ confirmed the previous findings, while those at 160 and $200\,\mu\mathrm{g\,m}^{-3}$ showed lesser changes that also become progressively greater with duration of exposure [30].

Thus, it is now clear that the appropriate averaging time for transient functional decrements caused by O_3 is ≥ 6 h, and there is no scientific basis for a health based exposure limit with an averaging time of 1 h. Since O_3 exposures in ambient air can have broad peaks with 8 h averages equal to $\sim 90\%$ of the peak 1 h averages, the functional decrements associated with ambient concentrations are likely to be much greater than those predicted on the basis of the responses in the chamber studies following 1–2 h exposures.

Symptomatic Responses

Respiratory symptoms have been closely associated with group mean pulmonary function changes in adults acutely exposed in controlled exposures to O_3 and in ambient air containing O_3 as the predominant pollutant. However, Hayes et al. [31] found only a weak-to-moderate correlation between FEV_1 changes and symptoms severity when the analysis is conducted using individual data.

In controlled 2 h O_3 exposures, McDonnell et al. [22] reported that some heavily exercising adult subjects experienced cough, shortness of breath, and pain on deep inspiration at $240\,\mu\mathrm{g\,m}^{-3}$, although the group mean response was statistically significant for cough only. Above $240\,\mu\mathrm{g\,m}^{-3}$ O_3, respiratory and non-respiratory symptoms have included throat dryness, chest tightness, substernal pain, cough, wheeze, pain or deep inspiration, shortness of breath, dyspnea, lassitude, malaise, headache, and nausea.

The recent prolonged exposure studies involving 6.6 h of exposure at concentrations between 160 and $240\,\mu\mathrm{g\,m}^{-3}$ also produced significant increases in respiratory symptoms including cough and pain on deep inspiration [29, 30]. Linder et al. [32] reported that brief exposures (16–28 min) at high ventilatory rates (30–120 L/min) produced symptoms of irritation and cough in young adults for exposures at 240 to $260\,\mu\mathrm{g\,m}^{-3}$.

While O_3 causes symptomatic responses in adults at current peak levels, there is a large set of data indicating that such responses do not occur in healthy children [33, 34]. While children (ages 8–11) exposed for 2.5 h at 240 μg m^{-3} O_3 while intermittently exercising ($\dot{V}_E$ = 39 L/min) showed small, but statistically significant decreases in FEV$_1$, they showed no changes in frequency or severity of cough compared to control [6, 35]. Similarly, adolescents (age 12–15) continuously exercising ($\dot{V}_E$ = 31–33 L/min) during exposure to 280 μg m^{-3} mean O_3 in ambient air showed no changes in symptoms despite statistically significant decrements in group mean FEV$_1$ (4%), which persisted at least one hour during resting post-exposure [33].

These laboratory results are consistent with the results obtained in a series of field studies at childrens' summer camps which failed to find any symptomatic responses despite the occurrence of relatively large decrements in function which were proportional to the ambient O_3 concentrations [24, 36].

Several epidemiology studies have provided evidence of qualitative associations between ambient oxidant levels > 200 μg m^{-3} and symptoms in children and young adults, such as throat irritation, chest discomfort, cough, and headache. Thus, symptoms reported in individuals exposed to O_3 in purified air are similar to those found for ambient air exposures, except for eye irritation, a common symptom associated with exposure to photochemical oxidants, which has not been reported for controlled exposures to O_3 alone. Other oxidants, such as aldehydes and peroxyacetyl nitrate (PAN), are primarily responsible for eye irritation and are generally found in atmospheres containing higher ambient O_3 levels [37, 38].

Respiratory symptoms in healthy young women (100 student nurses) in Los Angeles, in relation to ambient pollution levels, were monitored by Hammer et al. [39]. They found associations between photochemical oxidants and respiratory symptoms. However, that analysis ignored smoking and serial correlation in the data, used linear regression to model the probability of a respiratory incident, ignored other potentially colinear air pollutants, and assumed a pollutant would have the same impact on starting an episode of symptoms as on prolonging the episode. Because of limited computational facilities at the time, the data on individual subjects were collapsed to rates per day.

Schwartz and Zeger [40] reexamined the original diaries from this study containing smoking and allergy histories as well as symptom reports that had never been analyzed. Diaries were completed daily and collected weekly for as long as 3 years. Air pollution was measured at a monitoring location within 4 km of the school. Incidence and duration of a symptom were modeled separately. Photochemical oxidants were associated with increased risk of chest discomfort (OR = 1.7 p < 0.001) and eye irritation (OR = 1.20 p < 0.001).

Respiratory Disease

There have been several studies reporting associations between ambient photochemical oxidant pollution and exacerbation of asthma [41–43], but the role of O_3

specifically and the nature of the exposure-response relationships remain poorly defined.

Associations between ambient air pollutants and respiratory morbidity were examined by Ostro and Rothschild [44] using the Health Interview Survey (HIS), a large cross-sectional database collected by the National Center for Health Statistics. They attempted to determine the separate health consequences of O_3 and particulate matter, using six separate years of the HIS. The results, using a fixed effects model that controls for intercity differences, indicate an association between fine particulate and both minor restrictions in activity and respiratory conditions severe enough to result in work loss and bed disability in adults. Ozone, on the other hand, was associated only with the more minor restrictions.

Bates and Sizto [45] examined associations between ambient air pollutants and hospital admissions for respiratory disease in Southern Ontario. They found a consistent association in summer between hospital admissions for respiratory disease and daily levels of $SO_4^=$, O_3, and temperature, but no association for a group of nonrespiratory conditions. Multiple regression analyses showed that all environmental variables together accounted for 5.6% of the variability in respiratory admissions, and that when temperature was forced into the analysis first, it accounted for only 0.89% of the variability. It was found that daily $SO_4^=$ data collected at one monitoring site in the center of the region were not correlated with respiratory admissions, whereas the $SO_4^=$ values collected every sixth day, on different days of the week, at 17 stations in the region had the highest correlation with respiratory admissions. They concluded that probably neither O_3 nor $SO_4^=$ alone is responsible for the observed associations with acute respiratory admissions, but that either some unmeasured species (of which H_2SO_4 is the strongest candidate), or some pattern of sequential or cumulative exposure was responsible for the observed morbidity.

Another study of hospital admissions in relation to O_3 was performed by Ozkaynak, Kinney and Burbank [46] for Boston, Springfield, Worcester, New Bedford and Fall River, MA. They reported that the associations found between daily air pollution levels and admissions due to certain respiratory diagnosis classes were not always positive or significant, perhaps due to sample size limitations or other statistical problems. Nevertheless, the most complete and statistically reliable models indicated positive associations between 1-h maximum O_3 levels in the summer months and the daily admissions for pneumonia and influenza. Thus, in this respect these findings were similar to those obtained by Bates and Sizto [45].

Effects on Lung Defenses and Lung Structure

Laboratory Studies

Practical and ethical considerations limit the amounts and kinds of data that can be collected on the effects of O_3 on lung defenses and lung structure. Most of the limited body of data on humans that are available relate to the rate of particle clearance from the lungs and to alterations in the constituents of bronchoalveolar lavage (BAL).

Foster et al. [47] studied the effect of 2 h exposures to 400 or 800 $\mu g\,m^{-3}$ O_3 with intermittent light exercise on the rates of tracheobronchial mucociliary particle clearance in healthy adult males. The 800 $\mu g\,m^{-3}$ O_3 exposure produced a marked acceleration in particle clearance from both central and peripheral airways, as well as a 12% drop in FVC. It is of interest that the 400 $\mu g\,m^{-3}$ O_3 exposure produced a significant acceleration of particle clearance in peripheral airways, but failed to produce a significant reduction in FVC, suggesting that significant changes in the ability of the deep lung to clear deposited particles take place before significant changes in respiratory function take place.

The effects of O_3 on mucociliary particle clearance have also been studied in rats and rabbits. Rats exposed for 4 h to O_3 at concentrations in the range of 800 to 2400 $\mu g\,m^{-3}$ exhibited a slowing of particle clearance at ≥ 1600 $\mu g\,m^{-3}$, but not at 800 $\mu g\,m^{-3}$ [48]. Rabbits exposed for 2 h at 200, 500, and 1200 $\mu g\,m^{-3}$ O_3 showed a concentration dependent trend of reduced clearance rate with increasing concentrations, with the change at 1200 $\mu g\,m^{-3}$ being $\sim 50\%$ and significantly different from control [49]. It is not known why the animal tests show only retarded mucociliary clearance in response to O_3 exposure, while the human tests show accelerated clearance. In corresponding tests with other irritants, i.e. H_2SO_4 aerosol and cigarette smoke, both humans and animals have exhibited accelerated clearance at lower exposures and retarded clearance at higher exposures [50].

Studies of the effects of O_3 on alveolar macrophage mediated particle clearance during the first few weeks have also been performed in rats and rabbits. Rats exposed for 4 h to 1600 $\mu g\,m^{-3}$ O_3 had accelerated particle clearance [48]. Rabbits exposed to 200, 1200, or 2400 $\mu g\,m^{-3}$ O_3 once for 2 h had accelerated clearance at 200 $\mu g\,m^{-3}$ and retarded clearance at 2400 $\mu g\,m^{-3}$. Rabbits exposed for 2 h d^{-1} for 13 consecutive days at 200 or 1200 $\mu g\,m^{-3}$ O_3 had accelerated clearance for the first 10 d, with a greater effect at 1200 $\mu g\,m^{-3}$ [51].

Kehrl et al. [52] have studied the effects of inhaled O_3 on respiratory epithelial permeability in healthy, nonsmoking young men. They were exposed for 2 h to purified air and 800 $\mu g\,m^{-3}$ ozone while performing intermittent high intensity treadmill exercise. Specific airway resistance (SR_{aw}) and FVC were measured before and at the end of exposures. Seventy-five minutes after the exposures, the pulmonary clearance of a radioisotope labelled organic molecule, i.e., diethylene triamine pentacetic acid (^{99m}Tc-DTPA) was measured as an index of epithelial permeability. O_3 exposure caused repiratory symptoms in all 8 subjects and was associated with a $14 \pm 2.8\%$ (mean $\pm$ S.E.) decrement in FVC (p < 0.001) and a $71 \pm 22\%$ increase in SR_{aw} (p $= 0.04$). Compared with the air exposure day, 7 of the 8 subjects showed increased ^{99m}Tc-DTPA clearance after the O_3 exposure, with the mean value increasing from 0.59 ± 0.08 to $1.75 \pm 0.43\%$/min (p $= 0.03$). Thus, O_3 exposure sufficient to produce decrements in the respiratory function of human subjects also causes an increase in permeability. An increased permeability could facilitate the uptake of other inhaled toxicants and/or the release of inflammatory cells such as neutrophils onto the airway surfaces.

Seltzer et al. [53] showed that O_3-induced airway reactivity to methacholine is associated with polymorphonuclear leukocytes (PMN) influx into the airways and with changes in cyclooxygenase metabolites of arachidonic acid. For 2 h exposures to O_3 at $800 \, \mu g \, m^{-3}$ with intermittent exercise, the BAL fluid had increased prostaglandins E_2 and $F_{2\alpha}$, and Thromboxane B_2 at 3 h after the O_3 exposure.

Reports by Koren et al. [54, 55] also described inflammatory and biochemical changes in the airways following O_3 exposure. In their initial studies, subjects were exposed to $800 \, \mu g \, m^{-3}$ for 2 h while performing intermittent exercise at a ventilation of 70 L/min in order to examine cellular and biochemical responses in the airways. BAL was performed 18 h after the O_3 exposure. An $8.2 \times$ increase in PMNs was observed after ozone exposure, confirming the observations of Seltzer et al. [53]. A twofold increase in protein, albumin, and IgG were indicative of increased epithelial permeability as previously suggested by the ^{99m}Tc-DTPA clearance studies of Kehrl et al. [52]. In addition to confirmation of these previous findings, Koren et al. [54] provided evidence of stimulation of fibrogenic processes.

Koren et al. [55] reported that an inflammatory response, as indicated by increased levels of PMN, was also observed in BAL fluid from subjects exposed to $200 \, \mu g \, m^{-3} \, O_3$ for 6.6 h. The 6.6 h at $200 \, \mu g \, m^{-3}$ produced a $4.8 \times$ increase in PMNs at 18 h after the exposure. Since the amount of O_3 inhaled in the $200 \, \mu g \, m^{-3}$ protocol was $\sim 2.5 \, \mu g$, while it was $\sim 3.6 \, \mu g$ in the $800 \, \mu g \, m^{-3}$ protocol, we might have expected a $(2.5/3.6) \times 8.2 = 5.7$ times increase in PMNs. The close correspondence of the observed to expected ratio suggests that lung inflammation from inhaled O_3 also has no threshold down to ambient background O_3 levels.

The above studies indicate that the inflammatory process caused by O_3 exposure is promptly initiated [53] and persists for at least 18 h [54]. The time course of this inflammatory response and the O_3 exposures necessary to initiate it, however, have not yet been fully elucidated. Furthermore, these studies demonstrate that cells and enzymes capable of causing damage to pulmonary tissues were increased and the proteins that play a role in the fibrotic and fibrinolytic processes were elevated as a result of ozone exposure.

Inflammatory reactions occur in the nasal passages as well as in the lungs. Graham and Koren [56] compared the cellular changes detected in nasal lavage (NL) with those detected in the bronchoalveolar lavage (BAL) taken from the same individual. This study demonstrated that PMN counts in the NL could be a useful, inexpensive means of studying the acute inflammatory effect of ozone and monitoring those effects in the lower lung.

The Overton and Miller [57] model of O_3 dosimetry within the lungs predicts similar airway deposition patterns for O_3 in rats and humans, with the greatest deposition in the vicinity of the respiratory acinus, i.e. the junction between the small conductive airways and the gas exchange region. A recent extension of this work, based upon differences in O_3 removal in the upper respiratory tract and fraction exhaled, suggests that humans have about twice the deposition rate at the respiratory acinus as rats [58]. Thus, the effects seen in the chronic animal inhalation studies are likely to be conservative estimates of the effects that occur in humans in areas with high chronic exposure, such as Southern California.

Most of the inhaled O_3 penetrates beyond the sites in the airways that trigger the functional responses. In this deeper region of the lung, at and just beyond the terminal bronchioles, the effects produced by O_3 include changes in biochemical indices, lung inflammation, and airway structure. Furthermore, the effects of O_3 exposure in this region appear to be cumulative and persistent, even in animals that have adapted to the exposure in terms of respiratory mechanics [21].

In groups of mice exposed to 400 $\mu g\, m^{-3}$ O_3 for 1, 3, or 6 h, superoxide anion radical production decreased 8, 18, and 35% respectively, indicating a progressive decrease in bacteriocidal capacity with increasing duration of exposure [59]. In a series of inhalation studies, rats were exposed to O_3 at constant concentrations of either 240 or 500 $\mu g\, m^{-3}$ for 12 h d^{-1} for 6 and 12 weeks, or to a daily cycle with a baseline of 120 $\mu g\, m^{-3}$ for 15 h with a broad peak for 8 h averaging 360 $\mu g\, m^{-3}$ for a period of 3 or 12 weeks. Huang et al. [60] found that hyperplasia of Type I alveolar cells in the proximal alveoli was linearly related to the cumulative O_3 exposure in the four groups. Thus, there is no apparent threshold for cumulative lung damage.

For some chronic effects, intermittent exposures can produce greater effects than those produced by a continuous exposure regime with higher cumulative exposures. For example, Tyler et al. [61] exposed two groups of 7 month old male monkeys to 500 $\mu g\, m^{-3}$ for 8 h d^{-1} either daily or, in the seasonal model, on days of alternate months during a total exposure period of 18 months. A control group breathed only filtered air. Monkeys from the seasonal exposure model, but not those exposed daily, had significantly increased total lung collagen content, chest wall compliance, and inspiratory capacity. All monkeys exposed to O_3 had respiratory bronchiolitis with significant increases in related morphometric parameters. The only significant morphometric difference between seasonal and daily groups was in the volume fraction of macrophages. Even though the seasonally exposed monkeys were exposed to the same concentration of O_3 for only half as many days, they had larger biochemical and physiological alterations and equivalent morphometric changes as those exposed daily. Lung growth was not completely normal in either exposed group. Thus, long-term effects of oxidant air pollutants that have a seasonal occurrence may be more dependent upon the sequence of polluted and clean air than on the total number of days of pollution, and estimations of the risks of human exposure to seasonal air pollutants from effects observed in animals exposed daily may underestimate long-term pulmonary damage.

Epidemiological Studies

Epidemiologic studies of populations living in Southern California suggest that chronic oxidant exposures do affect baseline respiratory function. Detels et al. [62] compared respiratory function at two points in time five years apart in Glendora (a high oxidant community) and in Lancaster (a lower oxidant community – but not low by national standards). Baseline function was lower in Glendora, and there was a greater rate of decline over 5 years. Table 1 shows a comparison of the annual change in lung function in Lancaster and Glendora from the Detels et al. [42] study with that reported for Tucson, AZ by Knudson et al. [63] for a comparable population of Caucasian non-smokers. The second highest 1 h O_3 concentrations in Tucson in all of 1981, 1982, and 1983 were 200, 240 and 220 μg m^{-3} [64]. In Lancaster there were 58 d in 1985 with 1 h O_3 maxima greater than 240 μg m^{-3} while in Azusa, adjacent to Glendora, there were 117 days in 1985 with 1 h O_3 maxima greater than 240 μg m^{-3} [65]. Thus, the three different rates of function decline in Table 1 appear to suggest an exposure-response relationship with potentially significant health importance. Kilburn et al. [66] reported that non-smoking and exsmoking wives of Long Beach shipyard workers had significantly lower values of FEV_1, mid expiratory flow, terminal expiratory flow, and carbon monoxide diffusing capacity than those in a matched population from Michigan. The oxidant exposures in Long Beach and Michigan are not known, but those in Long Beach are similar to those in Lancaster, while those in Michigan are generally much lower. Both of these epidemiological studies have some serious methodologic deficiencies, but deserve citation in this discussion because they suggest effects that are consistent with the findings in the chronic animal exposure studies, i.e. they suggest premature aging of the lung in terms of lung function which might be expected on the basis of the cumulative changes in lung structure seen in the animals undergoing chronic exposure protocols.

Further evidence for chronic effects of O_3 were recently reported by Schwartz [67] based upon an analysis of pulmonary function data in a national population study in 1976–80, i.e. the second National Health and Nutrition Examination

Table 1. Annual change in lung function

Population (number)	FEV$_1$ (ml)	FVC (ml)	FEF$_{25-75\%}$ (ml/sec)	$\dot{V}_{50}$ (ml/sec)	$\dot{V}_{75}$ (ml/sec)	O_3 > NAAQS (No. d/yr)
Males						
Tucson (86)[a]	− 29	− 30	− 36	− 37	− 23	~ 1[d]
Lancaster (153)[b]	− 46	− 51	− 47	− 65	− 44	58[e]
Glendora (168)[b]	− 48	− 60	− 89	− 112	− 69	117[f]
Females						
Tucson (176)[c]	− 19	− 17	− 31	− 24	− 25	~ 1[d]
Lancaster (286)[b]	− 33	− 38	− 53	− 77	− 41	58[e]
Glendora (325)[b]	− 44	− 44	− 97	− 109	− 76	117[f]

[a] White, non-Mexican-American, > 25 yr of age who never smoked (Knudson et al., *AARD* 127: 725 1983).

[b] White, non-Spanish surnames only, 19–59 yr who never smoked. Test results between baseline and retest 5 yr later. Had not changed job or residence because of a respiratory problem. FEV$_1$ exists at both times (Detels et al., *Chest* 92: 594, 1987).

[c] White, non-Mexican-American, > 20, but < 70 yr of age who never smoked (Knudson et al., *ARRD* 127: 725, 1983).

[d] Second highest 1 h O_3 for 1981, 1982, and 1983 were 0.10, 0.12, and 0.11 ppm (EPA-ECAO CD for O_3 – 1986).

[e] CAL-ARB-Effects of O_3 on Health-Tech. Support Doc't (1987) – 1985 data.

[f] CAL-ARB – 1985 data for Azusa (~ 3 mi from Glendora).

Survey (NHANES II). Using ambient O_3 data from nearby monitoring sites, he reported a highly significant O_3 associated reduction in lung function for people living in areas where the annual average O_3 concentrations exceeded 80 $\mu g\,m^{-3}$.

Effects of Other Pollutants on Responses to Ozone

An important recent study which addressed the issue of the potentiation of the characteristic functional response to inhaled O_3 by other environmental cofactors was performed in Tuxedo, NY [36]. It involved healthy adult nonsmokers engaged in a daily program of outdoor exercise with exposures to an ambient mixture containing low concentrations of acidic aerosols and NO_2 as well as O_3. Each subject did the same exercise each day, but exercise intensity and duration varied widely between subjects, with minute ventilation ranging from 20 to 153 L, with an average of 79 L, and with duration of daily exercise ranging from 15 to 55 min, with an average of 29 min. Respiratory function measurements were performed immediately before and after each exercise period. O_3 concentrations during exercise ranged from 42 to 250 $\mu g\,m^{-3}$. All measured functional indices showed significant (p < 0.01) O_3 associated mean decrements. As shown in Table 2, the functional decrements were similar, in proportion to lung volume, to those seen in children engaged in supervised recreational programs in summer camps. They were as large (FEV_1) or much larger (FVC, FEF_{25-75}, PEFR) than those seen in controlled 1 and 2 h exposures in chambers. For the subgroup (7M, 3F) with the most comparable levels of physical activity, the responses in the field study were even greater. Since the ambient exposures of the adults exercising out of doors were for ~ 1/2 h, as compared to the 1 or 2 h exposures in the chamber studies, it was concluded that ambient cofactors potentiate the responses to O_3. Thus, the results of the exposures in chambers to O_3 in purified air underestimate the O_3 associated responses which occur among populations engaged in normal outdoor recreational activity and exposed to O_3 in ambient air.

The study on exercising adults, and earlier studies on children at summer camps [24, 68] were not able to demonstrate the specific effect of any of the measured environmental variables, including heat stress and acid aerosol concentration, on the O_3-associated responses. The inability to show the individual effects of other environmental cofactors on the response to ambient O_3 may be due to inadequate knowledge on the appropriate biological averaging time for these other factors. However, in the study of functional responses of children to ambient pollution in Mendham, NJ, a weeklong baseline shift in peak flow rate (PEFR) was associated with both O_3 and H_2SO_4 exposures during a four day pollution episode that preceded it [68]. A similar response to a brief episode with elevated O_3 and a much higher peak 4 h concentration of H_2SO_4 (46 $\mu g\,m^{-3}$) was seen among girls attending a summer camp in 1986 at Dunnville, Ontario, Canada, on the northeast shore of Lake Erie [69].

Controlled human exposure studies in chambers have not demonstrated synergism in functional response between O_3 and NO_2 or H_2SO_4, although Stacy et

Table 2. Mean functional changes per ppb O_3 after moderate or heavy exercise – comparison of results from field and chamber exposure studies with $O_3 \leq 180$ ppb

Study-Subjects, Age	Min. Vent. (liters)	Exposure (Exercise) Time (min)	O_3 Conc. (ppb)	Mean Rate of Functional Change		
				FVC (ml/ppb)	FEV_1 (ml/ppb)	FEF_{25-75} (ml/s/ppb)
Folinsbee (1988) 10M, 18–33 yrs	40	395(300)	120[b]	− 3.8	− 4.5	− 5.0
Gibbons (1984) 10F, 22.9 ± 2.5*	55	60(60)	150[b]	− 1.1	− 1.0	− 0.6
Avol (1984) 42M, 8F, 26.4 ± 6.9*	57	60(60)	153[a] 160[b]	− 1.2 − 1.5	− 1.3 − 1.5	– –
McDonnell (1983) 22M, 22.3 ± 3.1*	65	120(60)	120[b]	− 1.4	− 1.3	− 2.9
20M, 23.3 ± 2.0*	65	120(60)	180[b]	− 1.8	− 1.6	− 3.0
Kulle (1985) 20M, 25.3 ± 4.1*	68	120(60)	150[b]	− 0.5	− 0.2	− 2.1
Linn (1986) 24M, 18–33	68	120(60)	160[b]	− 0.7	− 0.6	− 1.1
Spektor (1988) 1M, 9F, 28–44	38.4 ± 12.3*	34.4 ± 9.9*	21–124[a]	− 1.9	− 1.8	− 6.7
7M, 3F, 22–40	64.6 ± 10.0*	26.7 ± 8.7*	21–124[a]	− 2.9	− 3.0	− 9.7
Lippmann (1983) 34M, 24F, 8–13	–	150–550	46–110[a]	− 1.1	− 0.8	–
Spektor (1988) 53M, 38F, 7–13	–	150–550	19–113[a]	− 1.0	− 1.4	− 2.5
Avol (1987) 33M, 33F, 8–11	22	60(60)	113[a]	− 0.3	− 0.3	–
Avol (1985) 46M, 13F, 12–15	32	60(60)	150[b]	− 0.7	− 0.8	− 0.7
McDonnell (1985) 23M, 8–11	39	150(60)	120[b]	− 0.3	− 0.5	− 0.6

*– Mean ± S.D.
[a] – Ozone concentration within ambient mixture
[b] – Ozone concentration within purified air

al. [70] did report that the mean responses to 800 $\mu g\,m^{-3}$ O_3 and 100 $\mu g\,m^{-3}$ H_2SO_4 after 2 h of exposure at rest were − 9.0% for FVC and − 11.5% for FEV_1, compared to corresponding values of − 5.7 and − 7.7% for O_3 alone, − 1.4 and − 1.2% for sham exposure, and + 0.9 and + 0.9% for H_2SO_4 alone. These mean differences appear to indicate an enhancement of the O_3 response by H_2SO_4. However, they were not statistically significant, perhaps because of the very high variability of the sham exposure results. Pollutant interactions that potentiate the characteristic O_3 response have been reported for other effects, in controlled exposure studies in animals, as will be discussed later in the sections on lung defenses and lung structure.

Exposure to O_3 can also alter the responsiveness of the airways to other bronchoconstrictive challenges as measured by changes in respiratory mechanics.

For example, Folinsbee et al. [29] reported that airway reactivity to the bronchoconstrictive drug methacholine for the group of subjects as a whole was approximately doubled following 6.6 h exposures to $240 \, \mu g \, m^{-3}$ O_3. Airway hyperresponsiveness (to histamine) had previously been demonstrated, but only at O_3 concentrations $\geq 800 \, \mu g \, m^{-3}$ [53, 71]. On an individual basis, Folinsbee et al. found no apparent relationship between the O_3-associated changes in methacholine reactivity and those in FVC or FEV_1. This differs from responses to inhaled H_2SO_4 aerosol, where changes in function correlated closely to changes in reactivity to carbachol aerosol, a bronchonconstrictive drug [72]. Perhaps the O_3 associated changes in bronchial reactivity predispose individuals to bronchospasm from other environmental agents such as acid aerosol and naturally occurring aeroallergens.

A variety of pollutant interactions that potentiate other characteristic O_3 responses have been reported in controlled exposure studies in animals. Last [73] reported synergistic interaction in rats, in terms of a significant increase in lung protein content, following 9 d exposures at $400 \, \mu g \, m^{-3}$ O_3 with 20 or $40 \, \mu g \, m^{-3}$ H_2SO_4, and a non-significant increase for 9 d at $400 \, \mu g \, m^{-3}$ O_3 with $5 \, \mu g \, m^{-3}$ H_2SO_4. Kleinman et al. [74] reported that lesions in the gas exchange region of the lung of rats exposed to O_3 were greater in size in rats exposed to mixtures of O_3 with either H_2SO_4 or NO_2. Graham et al. [75] reported a synergistic interaction between O_3 and NO_2 in terms of mortality in mice challenged with streptococcal infection either immediately or 18 h after pollutant exposure. Pinkerton et al. [76] reported that the clearance of asbestos fibers from the lungs of rats was reduced when the rats were also exposed to O_3. This increased fiber retention could increase the fibrogenic and carcinogenic risks of asbestos.

The ability of O_3 and other toxicants to act synergistically indicates that exposure limits for O_3 should include an extra margin of safety to acknowledge that co-exposures among ubiquitous pollutants such as O_3, NO_2, H_2SO_4, and asbestos are likely to occur in ambient air and many work environments.

Exposure-Response Relationships

In terms of the transient effects of O_3 on respiratory function in inidividuals, the effects appear to increase in proportion to the exposure durations up to six hours [29, 30] for O_3 concentrations down to $120 \, \mu g \, m^{-3}$ [24]. In terms of populations, there is a broad range of responsiveness among healthy young adults and children, and lesser responses among smokers, older people, and persons with asthma, allergic rhinitis, and chronic obstructive pulmonary disease. On the other hand, lesser responses among people with pre-existing lung disease may be more significant because of their reduced respiratory capacity [2].

In terms of chronic effects on baseline lung function [62] and structure [60], the dose-response relationships are less firmly established, but the data which are available suggest that these effects also increase in proportion to cumulative exposure.

Minimizing Effects from Ozone

The only two ways to reduce the effects of ambient O_3 are to reduce exposure and to reduce activity levels during times of peak exposures. Exposure levels are generally much lower indoors than outdoors since there are few significant indoor sources of O_3, and its high chemical reactivity causes rapid loss of outdoor O_3 infiltrating into indoor spaces. People exercising outdoors can reduce their O_3 uptake by doing the exercise before about 10 AM, when the outdoor O_3 concentrations begin their rapid rise.

Nitrogen Oxides

Introduction

There are a number of nitrogen oxides that can produce functional and structural effects in the respiratory tract. Among these, only nitrogen dioxide (NO_2) has an extensive health effects data base, albeit a still limited one. Nitric oxide (NO) can produce methemoglobemia, but only at levels well above current ambient air levels. Nitric acid (HNO_3) and nitrous acid (HNO_2) can be present at elevated levels, but we have little or no knowledge of their potential health effects. Ammonia (NH_3) from biogenic decay reacts with HNO_3 to form ammonium nitrate aerosol. This nearly neutral salt has not been demonstrated to produce health effects, but is a source for HNO_3 when present in airborne droplets that also contain hydrogen ion (H^+) from the dissociation of sulfuric acid (H_2SO_4).

Formation and Transport within the Atmosphere

NO and NO_2 are formed in fossil fuel combustion when nitrogen in the air or fuel combine with oxygen in the air at high temperature in the combustion zone. The ratio of NO to NO_2 varies with the combuster design and operation, and NO is gradually oxidized to NO_2 within the atmosphere downwind from the source. Concentrations in the air are usually reported as the sum of both, i.e. as NO_x. NO_2 in the atmosphere is gradually converted to nitrous acid (HNO_2), and nitric acid (HNO_3) vapors, especially at night.

Exposures to NO_2

The current U.S. NAAQS for NO_2 is $100\ \mu g\,m^{-3}$ (53 ppb) as an annual average, and nearly all U.S. communities have been in compliance with the NAAQS in recent years. On the other hand, millions of people are exposed to annual average concentrations which exceed the NAAQS because of the higher concentrations produced indoors by unvented combustion sources such as gas kitchen ranges and supplemental gas or kerosene space heaters, and the fact that most people spend most of their time indoors. Futhermore, there is much more temporal variation in indoor NO_2 than in outdoor NO_2, because of the patterns of usage of the indoor sources, raising concerns about the effects of peak as well as of average concentrations.

Where the current annual NO_2 standard of $100 \, \mu g \, m^{-3}$ is being met, 1 h max outdoor NO_2 concentrations exceed $300 \, \mu g \, m^{-3}$ on 10–20 d/year, except in California where they may exceed $300 \, \mu g \, m^{-3}$ on more than 40 d [77]. By contrast, indoor peak exposures can exceed $300 \, \mu g \, m^{-3}$ in kitchens with gas stoves during every major meal preparation.

Health Effects of NO_2 and HNO_3

The evidence cited as the primary basis of the original (1971) U.S. NAAQS for NO_2 of $100 \, \mu g \, m^{-3}$ was a group of epidemiology studies conducted in Chattanooga, which reported respiratory effects in children exposed to low level NO_2 concentrations over long-term periods. Reevaluation of the Chattanooga studies based on more recent information (especially regarding the accuracy of the air quality monitoring method for NO_2 used in the studies) indicates that these studies provide only limited evidence for an association between health effects in children and ambient exposures to NO_2.

The most frequent and significant NO_2-induced respiratory effects at concentrations of NO_2 below $4,000 \, \mu g \, m^{-3}$ (2 ppm) are: (1) lung function and symptomatic effects observed in controlled human exposure studies and in community epidemiological studies; (2) increased prevalence of acute respiratory illness and symptoms observed in outdoor community epidemiological studies and in indoor community epidemiological studies comparing residents of gas and electric stove homes; and (3) lung tissue damage, development of emphysema-like lesions in the lung, and increased susceptibility to infection observed in animal toxicology studies. The most recent NO_x Criteria Document concluded that results from these several kinds of studies collectively provided evidence indicating that certain human health effects may occur as a result of exposures to NO_2 concentrations at or approaching recorded ambient NO_2 levels. In the June 1985 promulgation of the reaffirmed NO_2 NAAQS (50 FR 25532), EPA identified young children and asthmatics as the groups at greatest risk from ambient NO_2 exposures. EPA believed that chronic bronchitics and individuals with emphysema or other chronic respiratory diseases may also be sensitive to NO_2 exposures. In addition, based on the findings from animal studies showing increased hematological, hormonal and other systemic alterations after exposure to NO_2, there was reason to believe that persons with cirrhosis of the liver or other liver, hormonal, and blood disorders, or persons undergoing certain types of drug therapies may also be more sensitive to

NO_2. Due to the lack of human experimental data for these latter groups, however, EPA considered the potential effects on such persons only as a factor in providing an adequate margin of safety.

The U.S. Bureau of the Census estimated that the total number of children under five years of age in 1970 was 17,163,000 and the total number between five and thirteen years was 36,575,000. Data from the U.S. National Health Survey for 1970 indicate that there were 6,526,000 chronic bronchitics, 6,031,000 asthmatics, and 1,313,000 emphysematics at the time of the Survey. Although there is overlap on the order of about one million persons for these last three categories, it is estimated that over twelve million persons experienced these chronic respiratory conditions in the U.S. in 1970.

Factors that EPA considered in assessing whether the current NO_2 standard provides an adequate margin of safety included: 1) Concern for potentially sensitive populations that have not been adequately tested; 2) concern for the effects of repeated peak exposures and delayed effects seen in animal studies but not yet examined in controlled human exposure studies; 3) possible synergistic or additive effects between NO_2 and other pollutants or environmental stresses; and 4) uncertainty about the NO_2 levels and duration of exposures associated with effects reported in the "gas stove" studies.

In recent years, there has been a substantial increase of our knowledge of the effects of low concentrations of NO_2 on lung function and structure. Bauer et al. [78] exposed asthmatic adults to $600\ \mu g\,m^{-3}$ (300 ppb) NO_2 for 20 min at rest followed by 10 min while exercising at a level that produced a 3-fold increase in minute ventilation. The subjects had greater reductions in FEV_1 after NO_2 than after sham exposure. They were also challenged with cold air, and prior NO_2 exposure potentiated the FEV_1, specific airway conductance, and airway reactivity responses to the cold air. Similar results were obtained by Roger et al. [79] for lung function. Mohsenin [80] exposed asthmatic adults at rest to $1,000\ \mu g\,m^{-3}$ (500 ppb) NO_2 for 1 h and reported a significant increase in airway reactivity, but no change in respiratory function. On the other hand, Linn et al. [81] found no effects of 600, 2,000 or $4,000\ \mu g\,m^{-3}$ on either respiratory function or airway reactivity in mildly asthmatic volunteers after 1 h of exposure with intermittent exercise. In normal subjects, Mohsenin [82] found that $4,000\ \mu g\,m^{-3}$ NO_2 affected airway reactivity.

It appears that HNO_3 can also affect respiratory function in asthmatic subjects, and that it is more potent than NO_2. Koenig et al. [83] exposed adolescent asthmatics for 30 min at rest followed by 10 min with exercise to $100\ \mu g\,m^{-3}$ HNO_3 and reported that it produced a significant decrease in FEV_1 in comparison to the sham exposure with air.

Low levels of NO_2 can also impair lung defenses against influenza virus. Smeglin et al. [84] exposed healthy adults to either sham air exposure to $1,200\ \mu g\,m^{-3}$ NO_2 for 3 h with intermittent exercise on 2 d. Alveolar macrophages harvested from bronchoalveolar lavage fluid collected at 3 h post-exposure were incubated for 2 d. There was significantly less viral inactivation by the macrophages exposed in vivo to NO_2 than those exposed in vivo to clean air, suggesting that such NO_2 exposures interfere with our ability to resist viral infections. Frampton et al. [85]

compared the protocol used by Smeglin et al. [84] with one involving the same total NO_2 exposure, but having three 15 min peaks of 4,000 $\mu g\,m^{-3}$ superimposed on a baseline level of 100 $\mu g\,m^{-3}$ for 3 h. The intermittent peak exposure did not produce the same reduction in viral inactivation as the constant 1,200 $\mu g\,m^{-3}$ exposure, suggesting that the base-line level is much more important than peak exposures. Kulle et al. [86] exposed healthy adult volunteers to 2,000 $\mu g\,m^{-3}$ for 2 h d^{-1} for 3 consecutive days, and administered a live, attenuated influenza virus intranasally after the exposure on the second day. They found that a greater percent of the subjects were infected after NO_2 exposure than after clean air exposure.

In chronic exposures of mice by Miller et al. [87], the effects of periodic spikes above a concentration baseline appeared to play a significant role in the effects produced. Three groups of animals received 52 weeks of 23 h d^{-1}, 7 d/week exposure to: a) clean air; b) 400 $\mu g\,m^{-3}$ NO_2; and c) 400 $\mu g\,m^{-3}$ NO_2 with two 1,600 $\mu g\,m^{-3}$ spikes for 1 h d^{-1} on 5 d/week. Both NO_2 exposed groups had significant treatment effects in terms of bacterial infectivity and pulmonary function, with the spiked exposure group having significantly greater responses.

The apparent discrepancy in the infuence of exposure spikes on health effects between the human acute studies of Frampton et al. [85] and the chronic mouse exposure studies of Miller et al. [87] could have been due to the differences in protocol, infective challenge, species, or other factors, such as the increment to the cumulative exposures of the mice by the peaks (6 ppm-h/week added to 32 ppm-h/week). Further research on the influence of spikes of exposure is clearly needed.

Minimizing Effects from NO_2

As with O_3, the effects of NO_2 can be minimized by reducing exposures and ventilatory rates. In the absence of unvented combustion sources such as gas cooking ranges, supplemental gas or kerosene fired space heaters, and tobacco smoking, indoor exposures will be somewhat lower than outdoor exposures. However, when such combustion sources are used indoors, indoor exposures can be substantially higher. Thus, to reduce exposures, one can either reduce the amount burned, or vent the combustion products to the outside. Outdoor NO_2 is attributed, in roughly equal measure, to motor vehicles and large stationary combusters, and reductions in exposure will require the application of new control technologies for such sources.

Sulfur Oxides

At the present time SO_2 is the index pollutant for the sulfur oxides, with both 24 h max. and annual average concentrations as primary standards. Most of the sulfur in fossil fuel is converted into SO_2 in the combustion zone and it is vented to the atmosphere with the other products of combustion. A small fraction of the sulfur, generally less than 10%, is emitted as sulfuric acid (H_2SO_4), with some of it being a surface film on ultrafine sized mineral ash particles. When the discharge point is a tall stack, most of the SO_2 escapes local deposition on terrestrial surfaces, and is gradually (1–10%/h) converted into SO_3, a highly hygroscopic vapor. The SO_3 rapidly combines with water vapor to produce an ultrafine droplet aerosol of H_2SO_4. The H_2SO_4 is then gradually neutralized by ammonia; first to the strong acid, ammonium bisulfate (NH_4HSO_4), and then to ammonium sulfate (($NH_4)_2SO_4$), a nearly neutral salt. Rates of ammonia neutralization vary widely, depending on emission sites from ground based sources. Rates are high over cities and agricultural areas, low over forests, and virtually nil over deep water bodies.

The ratios between SO_2, H_2SO_4, and total particulate sulfate ($SO_4^=$) in the atmosphere are highly variable in space and time. While ambient concentration data are relatively plentiful for SO_2 and $SO_4^=$, they are, unfortunately, very sparse for H_2SO_4 and NH_4HSO_4, the strong acid aerosols which are believed to account for most of the mortality and morbidity historically associated with mixtures of SO_2 and particulate matter (PM). Both SO_2 and $SO_4^=$ are very poor surrogate indices for ambient concentrations of acid aerosols.

Exposures to Sulfur Oxides

Current U.S. ambient air levels of SO_2 are generally well within the current primary U.S. Standard of 80 $\mu g\,m^{-3}$ for an annual average and 365 $\mu g\,m^{-3}$ for a 24 h maximum. There is an additional special concern for asthmatics' peak exposures to SO_2 while performing outdoor exercise. It has been estimated that the size of the asthmatic population with peak 5–10 min exposures at concentrations > 0.2 ppm (520 $\mu g\,m^{-3}$) during light to moderate exercise, who may exhibit a bronchoconstrictive response, varies from 5,000 to 50,000 in the U.S.

For acidic aerosols, there is a very limited ambient concentration data base. Lioy and Waldman [88] summarized the available data from research investigations conducted over limited time periods, generally during the summers. Levels of acidic aerosol in excess of 20–40 $\mu g\,m^{-3}$ (as H_2SO_4) have been observed for time durations ranging from 1 to 12 h. These were associated with high, but not necessarily the highest, atmospheric $SO_4^=$ levels. Exposures (concentration – time product) of 100 to 900 $\mu g\,m^{-3}$ h were calculated for the acid events which were monitored. In contrast, earlier London studies indicated that acidity in excess of 100 $\mu g\,m^{-3}$ h (as H_2SO_4) were present in the atmosphere, and exposures > 2000 $\mu g\,m^{-3}$ h were possible.

The first data on annual average acidic aerosol concentrations in contemporary U.S. communities were recently reported by Spengler et al. [26]. In the four eastern U.S. communities studies, the annual average ranged up to 1.8 $\mu g\,m^{-3}$ (as H_2SO_4).

Health Effects of SO$_2$

Asthmatics and "atopics" exhibit changes in pulmonary mechanics, such as increases in airway resistance, that are indicative of significant bronchoconstriction after exposures of only 5 to 10 min to SO$_2$ concentrations below 1,300 μg m^{-3}. At or above 1,300 μg m^{-3}, changes in functional measures are accompanied by perceptible symptoms such as wheezing, shortness of breath, and coughing. Available information indicates that these effects only occur with moderate (or greater) exercise levels and these sensitive individuals are about an order of magnitude more sensitive than "normal" subjects.

The magnitude of response (typically bronchoconstriction) induced by any given SO$_2$ concentration is variable among individual asthmatics. Exposures to SO$_2$ concentrations of 650 μg m^{-3} or less, which did not induce significant group mean increases in airway resistance, also did not cause symptomatic bronchoconstriction in individual asthmatics. On the other hand, exposures to 1,000 μg m^{-3} SO$_2$ or greater (combined with moderate to heavy exercise) which induced significant group mean increases in airway resistance, also caused substantial bronchoconstriction in some individual asthmatics. This bronchoconstriction was often associated with wheezing and the perception of respiratory distress. In a few instances it was necessary to discontinue the exposure and provide medication. The significance of these observations is that some SO$_2$-sensitive asthmatics are at risk of experiencing clinically significant (i.e. symptomatic) bronchoconstriction requiring termination of activity and/or medical intervention when exposed to SO$_2$ concentrations of 1,000 to 1,300 μg m^{-3} or greater when this exposure is accompanied by at least moderate activity. These are higher concentrations than the 24 h U.S. Standard of 365 μg m^{-3}, but can occur down wind of point sources as 10 min averages without exceedances of the 24 h Standard.

Health Effects of Acidic Aerosol (H$_2$SO$_4$ and NH$_4$HSO$_4$)

The human health effects of major concern with respect to the inhalation of acidic aerosol are bronchospasm in asthmatics and chronic bronchitis in all exposed persons. The former relates to acute exposure, while the latter can be related more closely to chronic or cumulative exposures. In either case, the effects are produced by droplets depositing on the surface of the conductive airways of the lungs.

The studies related to the provocation of bronchospasm show evidence for increased airway resistance in exercising mild to moderate asthmatics were reported by Koenig et al. [89, 90] following H$_2$SO$_4$ exposures for 30 min at rest and 10 min of exercise for adolescents at 68 or 100 μg m^{-3} (0.6 μm diameter droplets). Bauer et al. [91] saw similar effects in adults exposed for 2 h at 75 μg m^{-3} while exercising, or for 16 min at 450 μg m^{-3} (0.8 μm) while at rest. While the average increases in airway resistance were small, the populations studied were small, and some subjects had much greater responses than the average. Also, the populations were carefully selected and did not include the more unstable and potentially more reactive asthmatics in the population.

Spengler et al. [27] showed that the responses observed by Koenig in adolescent asthmatics exposed for 30 min at rest and for 10 min during exercise at 100 μg m^{-3} and those observed by Utell in adult asthmatics exposed at rest for 16 min at 450 μg m^{-3} are consistent when the total amounts of H_2SO_4 inhaled rather than the concentrations are considered. In the follow-up study, Koenig et al. [11] showed that the effects previously seen at 100 μg m^{-3}, following 30 min of resting exposure and 10 min of exposure with exercise, were also seen, albeit at an attenuated level, at 68 μg m^{-3}, and that these effects were increased when there was co-exposure to 260 μg m^{-3} SO_2. Schlesinger [92] studied the effects of concentration and duration of exposure to submicrometer sized H_2SO_4 aerosols on tracheobronchial and alveolar rates of particle clearance, and found that the effects increased with duration and concentration. Spektor et al. [93] reported a similar finding for tracheobronchial particle clearance in humans. Amdur and Chen [94] showed that sulfuric acid (at 20 μg m^{-3}) on ultrafine particles which simulate coal combustion effluent, when present in a mixture also containing SO_2, produce functional responses about 10 fold greater than those produced by pure droplets of H_2SO_4 of comparable size.

Thus, various studies indicate that the acute effects are more closely related to cumulative daily exposure than to peak concentrations, and that an exposure limit for H_2SO_4 should be expressed as a multihour average rather than as a short-term concentration limit.

With respect to concentration limits to protect against chronic bronchitis, a mechanistic basis for linkage between chronic exposure to H_2SO_4 and the pathogenesis of chronic bronchitis lies in the series of studies involving chronic exposures of animals and persistent histological alterations in lung structure. As noted by Lippmann et al. [95], these structural changes in the rabbit model have correlates in terms of clearance function changes. These, in turn, are indicative of changes in mucus secretion leading to mucus stasis, a hallmark of bronchitic disease.

The animal studies can be related to human responses in two ways. One is the concordance in functional and morphometric responses of animals to H_2SO_4 and cigarette smoke, a known causal factor for human chronic bronchitis. The other is that humans, rabbits, and donkeys all have essentially the same transient mucociliary clearance function responses to single 1 h exposures to H_2SO_4. The fact that daily one-hour H_2SO_4 exposures in rabbits and donkeys produce persistent changes in clearance function makes it highly likely that humans would also if similarly exposed. Furthermore, a comparison of the human and rabbit responses to single exposures indicates that humans respond at lower concentrations than do rabbits [96].

The effects of H_2SO_4 on the airways are very likely to be cumulative, during each exposure day, at least in part. Thus, the daily one-hour exposures at 250 μg m^{-3} in the rabbits may be equivalent to < 50 μg m^{-3} for a 7–8 h day and to a still lower concentration for equivalent effects in humans. On the other hand, the effects produced by the one-year series of exposures in the rabbits were less severe than the condition corresponding to a clinical diagnosis of chronic bronchitis in humans.

Several recent papers strengthen the basis for associating chronic inhalation exposures of ambient acidic aerosols with bronchitis in humans. The most direct evidence was presented by Speizer [97]. In five U.S. cities which varied in annual average ambient acid concentrations from 8 to 35 nmoles m^{-3} (0.4 to 1.8 μg m^{-3} of H_2SO_4 equivalent), the incidence of reported symptoms of bronchitis in the last year among 10–12 year old children varied in proportion to the ambient acid concentration (Fig. 2). Indirect, but supportive evidence was presented by Wichmann et al. [98]. During an acidic pollution episode in Jan. 1985 in the Ruhr district in West Germany in which average concentrations of SO_2 and suspended particles were 800 and 600 μg m^{-3}, there were significant increases in deaths, hospital admissions, outpatient visits, and ambulance deliveries to hospitals in comparison to those in a less polluted control area.

Dassen et al. [99] measured pulmonary function in primary school children in the Netherlands before, during, and after the Jan. 1985 acidic pollution event which caused elevated exposures throughout northwestern Europe. During this episode, the Dutch children were exposed to respirable particulate and SO_2 concentrations in the range of 200–500 μg m^{-3}. Their baseline respiratory functions had been measured about 4–6 weeks earlier, when the pollutants were below 100 μg m^{-3}. During the episode, function indices were significantly lower by 3–5%, and decrements were still present at 16 d later, but not at 25 d.

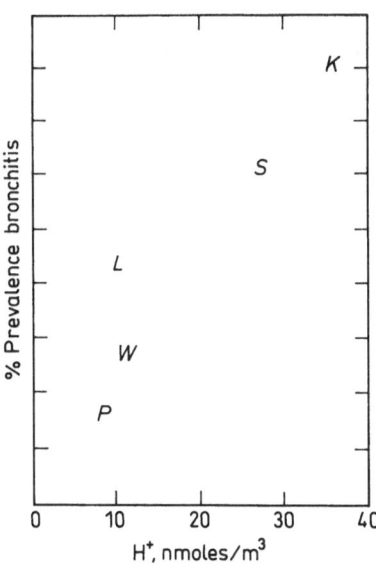

Fig. 2. Bronchitis in the last year in children 10 to 12 yrs of age in five U.S. cities by annual average concentration of hydrogen ion (P = Portage, WI; L = St. Louis, MO; S = Steubenville, OH; K = Kingston, TN; W = Watertown, MA). From: Speizer [97]

Minimizing Effects from Exposures to Sulfur Oxides

The most clearcut effect of SO_2 is a transient bronchoconstriction which occurs after about 5 min of exposure above about $520\ \mu g\,m^{-3}$ in asthmatic subjects performing vigorous exercise. Similar effects occur in healthy persons, but only at much higher SO_2 concentrations. Such effects can be minimized by avoiding vigorous exercise out-of-doors in locations within a few miles of a large power plant burning high sulfur coal or oil, or a smelter processing high sulfur ores, especially for stacks lacking SO_2 scrubbing. Asthmatics who persist in such outdoor exercise should carry a portable bronchodilator drug dispenser (inhaler) to control the bronchospasm which may occur from excessive exposures to SO_2.

Similar considerations apply to the acute transient effects of H_2SO_4 on respiratory function. A powerplant effluent in which 10% of the sulfur is emitted as H_2SO_4 could have $\sim 50\ \mu g\,m^{-3}$ of H_2SO_4 as well as $\sim 500\ \mu g\,m^{-3}$ of SO_2. Koenig et al. [90] reported that $68\ \mu g\,m^{-3}$ of H_2SO_4, combined with $260\ \mu g\,m^{-3}$ of SO_2, produced a greater effect than the $68\ \mu g\,m^{-3}$ of acid alone.

The effects on daily mortality rates and on respiratory function which persist for several weeks following pollution episodes lasting several days, and characterized by elevated concentrations of both SO_2 and PM (and presumably acid aerosol as well) can be minimized by restricting outdoor activity, and especially vigorous exercise out-of-doors during such episodes. Concentrations of SO_2 and PM during such episodes generally approach or exceed the current 24 h standards, and will be known to the public from advisories in the newspapers, radio and TV.

The effects on chronic respiratory symptoms, baseline lung function, and annual mortality rates which may be due to chronic exposures to acidic aerosols cannot readily be minimized by restrictions on personal activity or mobility. They can only be reduced by restricting the emissions of SO_2 and H_2SO_4 from large stationary sources.

Particulate Matter

Particulate matter is the only criteria pollutant with no definition of the regulated pollutant, other than the functional one of being matter suspended in air as solid or liquid particles. Thus, all manner of airborne materials contribute to the measured mass concentrations, including particulate materials dispersed into the air, such as suspended soil, resuspended road dust, combustion smoke and ash, sea salt, and process wastes from industrial operations. It also includes particulate matter formed within the atmosphere from gaseous precursors, such as H_2SO_4 and photochemical smog. The former category is called primary aerosol, and the latter is known as secondary aerosol.

Particulate matter is often subdivided into categories according to particle size. The smallest mode, known as the transient, or Aitken nuclei mode, includes particles less than ~ 0.1 μm in diameter. It forms in the atmosphere from gaseous precursors at low mass, but high particle number concentrations. The particles coagulate rapidly, with the aggregates growing larger than ~ 0.1 μm within about 10 min, passing over into the so-called accumulation mode (0.1 to ~ 2 μm diameter), where diameters and concentrations remain relatively stable over periods ranging from hours to days. The composition of the particles in the Aitken and accumulation modes is generally dominated by carbonaceous particles from combustion sources and photochemical reaction processes, and by sulfate particles. The latter are generally present as solution droplets whose diameters vary with ambient humidity.

Particles with aerodynamic diameters above 2 μm are generally known as coarse mode aerosols, and are composed largely of mechanically dispersed soil-like irregularly shaped particles. They are generally lighter in color, basic rather than acidic, and less persistent in the atmosphere due to their greater settling rates. Particles with aerodynamic diameters larger than ~ 10 μm are found only where active dispersion processes can continue to generate them. They are of little concern with respect to health effects since only a small proportion can penetrate through the oral or nasal passages to the lungs, where they can contribute to the lung diseases of concern. On this basis, the U.S. EPA changed the index pollutant for PM on July 1, 1987 from total suspended particulate matter (TSP) with an undefined upper size cut, to PM_{10}, which requires sampler inlets to remove more than 50% of particles greater than 10 μm in aerodynamic diameter (52 FR 24634).

Since particle composition is a major determinant of toxicity, it is difficult to justify mass concentration limits for particulate matter without reference to composition. This helps to account for the failure of PM limits to protect against effects due to acidic aerosol, a small, but highly variable mass constituent of PM. On the other hand, the enforcement of the PM NAAQS has almost certainly had beneficial effects on public health. The implementation of PM concentration reductions under the Clean Air Act has led to significant reductions in ambient concentrations of carbonaceous and metal oxide aerosols, many of which are toxic and/or carcinogenic. This provides a health based justification for retaining nonspecific mass concentration NAAQS, even if the promulgation of a new NAAQS for acidic aerosols takes place.

Exposures to Particulate Matter

Since most people spend most of their time indoors, their total exposure to PM depends, to a great extent, on the penetration into and persistence within the indoor air of PM from the outdoor air. It also depends on the nature and strength of indoor sources of PM. The largest source in homes or public buildings housing smokers is tobacco smoke. Other sources include unvented combustion sources for cooking and space heating, lint, resuspended dust from cleaning and maintenance, residues from consumer products, especially spray generated products, and air-borne products of hobby activities. Indoor concentrations of PM are generally lower than outdoor concentrations when smokers are absent, and generally higher when smokers are present.

Carbon Monoxide

Sources of CO

As an outdoor pollutant, carbon monoxide (CO) is almost entirely attributable to motor vehicle exhaust. Motor vehicle exhaust can also be a dominant source indoors, especially in residential units and public buildings with internal garages. Indoor concentrations in ice skating arenas can also be quite high from periodic use of ice scraping machines powered by internal combustion engines. Sidestream cigarette smoke can also contribute significantly to indoor CO concentrations. Unvented indoor combustion sources such as gas ranges and space heaters can cause very high concentrations of CO when they are not properly maintained or operated. However, they contribute relatively little CO under normal usage.

Exposures of CO

Outdoor concentrations exceeding the current U.S. Standard of 10 mg m^{-3} for 1 h max and 40 mg m^{-3} for 8 h max occur in the central cores of large U.S. metropolitan areas, due to the concentration of motor vehicles in confined street canyons. As a result, exposures of people above the NAAQS can be expected among people who spend a major fraction of their time out-of-doors in central cities peak traffic hours. Other people likely to have excessive exposure are those who work in or near parking garages, or those who live in urban areas at high altitudes such as Denver and Albuquerque, where the partial pressure of oxygen is reduced and the motor vehicle fuel is burned less completely.

Health Effects of CO

The oxygen deprivation mechanism of CO has been shown to have adverse effects on the human cardiovascular and central nervous systems, as well as on fetal development. The key effects identified by EPA to support the 1985 CO Standard were: (a) decreased work time to exhaustion in healthy young men occurs at 2.3–4.3% carboxyhemoglobin (COHb), (b) decreased duration of exercise before onset of pain in angina pectoris patients occurs at 2.9–4.5% COHb, (c) decreased minimal oxygen consumption and exercise time during strenuous exercise in healthy young men occurs at 4.0–20% COHb, (d) vigilance decrements do not occur below 5.0% COHb, but do occur between 5.0 and 7.6% COHb, and (e) diminution of visual perception, manual dexterity, ability to learn or performance on complex sensorimotor tasks (such as driving) occurs at 5.0–17% COHb.

Based on this health effects evidence, consideration of uncertainty in the Coburn model (which relates CO exposure to COHb levels), and exposure analysis estimates from the NAAQS Exposure Model (NEM), the EPA decided that the 8 h, $10 \, mg \, m^{-3}$ standard would keep more than 99.9% of the nonsmoking cardiovascular population below 2.1% COHb.

To help resolve some of the remaining uncertainties about the CO exposure-effects relationships, the Health Effects Institute supported a multi-center study of CO effects in angina patients [100], similar to earlier studies on which the Standard for CO was based. Specifically, the purpose of this study was to determine whether or not exposures to CO for 50–70 min at 117 and 253 ppm (producing approximately 2% or 4% COHb levels respectively) cause an exacerbation of myocardial ischemia in subjects with coronary artery disease during a progressive exercise test. The study was performed at Johns Hopkins University School of Medicine, Rancho Los Amigos Medical Center and St. Louis University School of Medicine as a double-blind, placebo-controlled research effort. Exposures to air or CO were randomized within each center.

For enrollment in the study, the men had to meet stringent entry criteria. They had to be nonsmokers, between 35 and 75 years of age, and have stable, reproducible, exertional angina pectoris and positive exercise treadmill tests with reproducible ST-segment changes. In addition, the men had to meet at least one of three objective indicators of coronary artery disease: (1) angiographic evidence of 70% or more narrowing of at least one major coronary artery; (2) previously documented myocardial infarction; or (3) a positive exercise thallium test. The enrolled subjects were asked to remain on their normal dose and timing of medication.

On the first visit to the laboratory, each subject underwent a physical examination and was screened with baseline blood tests, including COHb levels. In order to ensure that the exercise-induced ST-segment changes and angina were reproducible, the subject twice performed a standardized treadmill test, with a sham exposure period between the two tests. Subsequently, he was exposed to 150 ppm CO while at rest, and his rate of CO uptake was determined.

The subjects who met all entry criteria were asked to return for three additional visits. The test visits were identical, except that subjects were randomly assigned to exposures of either fresh air or one of the two levels of CO. During each test visit, each subject twice performed a symptom-limited standardized exercise treadmill test, with a recovery and exposure period between the two tests.

The two primary health endpoints used in this study were (1) the time to onset of exercise-induced angina, and (2) the time to predefined ST-segment change. Unprocessed and computer-processed ECG records were analyzed and evaluated by two cardiologists who were blinded to the subject's identification and exposure level. The ST-segment change provided an objective measure of cardiac ischemia, whereas the angina endpoint was necessarily subjective. The total duration of exercise, the duration of the ST-segment change, the amplitude of maximum ST change, and the double product (heart rate multiplied by blood pressure) at the time of ST change and the angina were also examined as secondary endpoints.

Based upon an analysis of the 63 subjects who met all of the protocol criteria, the investigators concluded that at both the 2% and 4% COHb target levels, there were statistically significant decrements in the time to onset of ST-segment change, and the times to onset of angina, compared to air exposure. At an average COHb level of 2% (representing an increment of 1.4 from the average baseline level of 0.6% COHb), there was an average 5.1% reduction in time to ST-segment change, while for 4% COHb, there was an average 12.1% reduction. The reductions in time to onset of angina at the same levels of COHb were 4.2% and 7.1%, respectively.

The positive findings from the primary endpoint analyses were further supported by significant differences in some of the secondary endpoints. Maximum ST amplitude and the summed ST score increased at both levels of CO exposure, while the heart rate-blood pressure double product at the time of ST change, and the total exercise duration, were significantly different only at the 4% COHb target level.

Overall, the results of this study suggest that cardiovascular patients should avoid exposures that approach the current ambient air quality standard for CO.

Lead

Sources of Lead Exposure

For infants and small children, significant body burdens of lead can be acquired from ingested soil and paint chips. Household dust and garden soil, especially in urban areas can be greatly enriched in lead from the fallout of airborne particles from motor vehicles burning leaded fuels. Another major source in older homes is paint flakes and chips containing lead-based pigments. It has been illegal to use such pigments in interior paint for more than fifty years, but there are large inventories within older buildings which can be readily mobilized and dispersed during maintenance and renovations. In 1988, ATSDR [101] indicates that 40 million households in the U.S. contain hazardous quantities of leaded paint.

Yaffe et al. [102] reported that the isotopic ratios of lead in the blood of children were close to the average lead ratios of paints from exterior walls and to the lead ratios of surface soils in adjacent areas where the children played. Their data suggest that the lead in the soil was derived mainly from weathering of lead-based exterior paints, and that the lead-contaminated soil was a proximate source of lead in the blood of the children.

As shown in Fig. 3, the most significant contributors to current body burdens of lead include direct air inhalation, inhalation or direct ingestion of settled dust, and

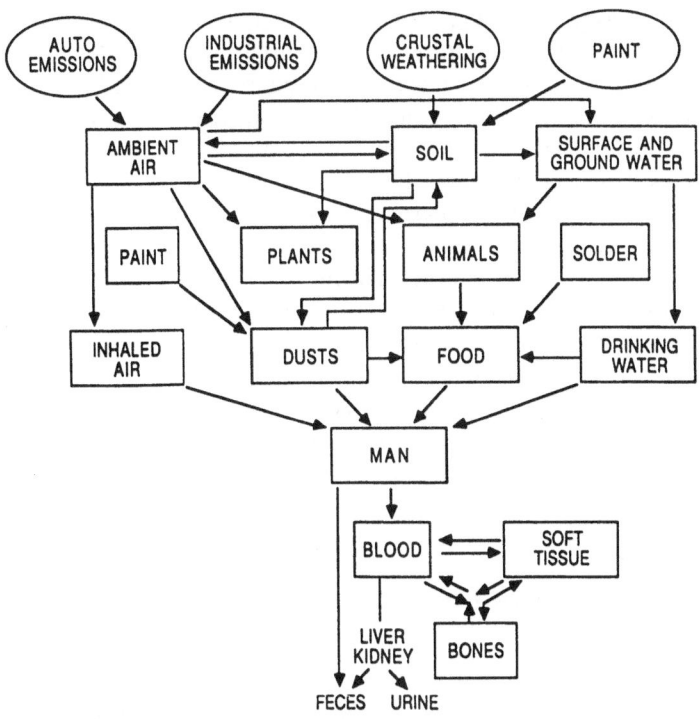

Fig. 3. Pathways of lead from the environment to and within man (EPA, 1986)

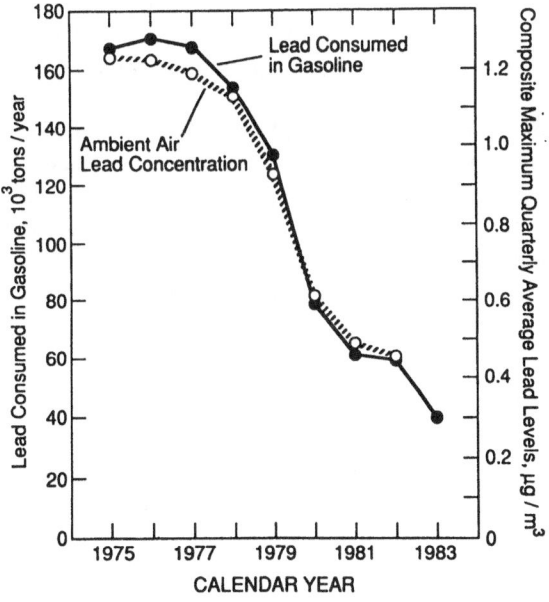

Fig. 4. Lead consumed in gasoline and ambient lead concentrations, 1975–1983 (EPA, 1986)

the ingestion of food and water. Some old housing stock has lead pipe that can elevate potable water concentrations substantially. Lesser, but still significant elevations can occur in water delivered via modern copper and brass pipe due to leaching of lead from the solder in the joints. Foods can be enriched in lead from a variety of sources. Lead in the air can deposit on leafy vegetables and fruits and leave residues which are ingested. Lead in the soil can be incorporated into the growing plant. Canned foods can also extract lead from solder used to seal the can.

Lead exposures and blood lead levels among the general population have declined substantially in recent years. The most substantial reduction has been due to the switch from leaded to unleaded gasoline as motor vehicle fuel. This had an immediate effect on air lead (Fig. 4) and a parallel reduction in average blood lead (PbB) concentration that lagged by $\sim$ 3 months. The lag occurred because most of the tailpipe lead reached people indirectly through exposure to resuspended soil and through incorporation into foodstuffs. Further reductions have occurred as the food packaging industry has reduced the use of solder in cans.

Recent Health Effects Findings

The literature on human exposures to lead and their health effects is voluminous. The 1986 EPA Criteria Document was published in four volumes containing 1,336 pages. This discussion will be limited to the more descriptive research relative to low-level population exposures and the chronic health effects associated with such exposures. In most cases, this limits the review to studies of the associations between exposure and health effects in humans, since the low-dose effects of interest have seldom been seen in animal toxicology studies conducted at higher levels of exposure.

The effects that have been associated with blood lead (PbB) concentrations $< 40\ \mu g\,dL^{-1}$ will be the main focus of this selective review. These include the effects of prenatal and early childhood exposures on physical and neurobehavioral development of children, and the influence of chronic low-level exposure on cardiovascular function in adults. These low-exposure related effects are of interest in relation to both occupational and general environmental exposures.

Neurobehavioral and Developmental Effects in Children

While various maladaptive behaviors, neuropsychological deficits, and neuro-anatomical changes have been associated with chronic exposures to relatively low concentrations of lead, no single mechanism appears sufficient to account for the diverse effects. It is more likely that lead acts at several cellular and sub-cellular sites. Lead readily enters the brain and appears to be selectively deposited in the hippocampus and cortex as well as in non-neuronal elements that are important in the maintenance of "blood-brain barrier" functions. Once deposited, lead is retained in the brain for long periods of time even after external exposure ceases and PbB levels decline. These spatial and temporal patterns of brain lead accumulation correspond to neurobehavioral and morphologic abnormalities associated with lead exposure. The sensitivity of the brain during the period of maximal brain growth and differentiation in the first 2 years of life tends to magnify the severity of the long-term consequences.

Low PbB levels may contribute to behavioral disorders, such as attentional deficits and distractibility in essentially normal children not diagnosed as hyperactive. A study by Bellinger et al. [103] suggests that measures of classroom performance may show long-term effects of early lead exposure. Winneke et al. [104] found that behavioral and attentional deficits as rated by teachers (e.g. disordered classroom activity, restless, easily distracted, not persistent, does not follow directions, low overall functioning) were significantly associated with children's tooth and PbB levels, which was consistent with the earlier association reported by Needleman et al. [105].

In addition, lead levels in young children have been consistently associated, following appropriate adjustments, with deficits in reaction time under varying intervals, which is an index of attentiveness, and with reaction behavior. These findings argue for probable effects of lead on attention and vigilance functions at PbB levels extending below $30\,\mu g/dL$, and possibly, down to as low as 15–$20\,\mu g\,dL^{-1}$.

There is also evidence that low levels of lead may be associated with effects on some complex cognitive functions including learning, visual-perception skills, and IQ scores. The studies on children have attracted controversy because of difficulties associated with attributing subtle deficits in child development to lead exposure rather than to effects due to genetics, nutrition, medical history, access to education, and parental and social influences, all of which interact in potentially complex ways to mold an individual.

Considerable uncertainty has existed regarding lead's impact on IQ scores of children with PbB levels below $40\,\mu g\,dL^{-1}$. This uncertainty stems largely from the complex interaction between lead exposure over time, social factors, and intelligence scores, from the statistical and methodological limitations of cross-sectional studies to untangle these variables, and the range of interpretations that result from these studies.

From the Needleman et al. [105] study, and subsequent re-analyses that controlled for confounding variables including pica [106], there were average IQ decrements of about 4 points and other neurobehavioral deficits that appeared to be associated with lead exposures of U.S. children that resulted in dentine lead values that exceeded 20–$30\,ppm$ (indicating average PbB levels in the 30–$50\,\mu g\,dL^{-1}$ range). Needleman et al. [107] calculated that a 4 point decrement in the mean IQ of a normal population distribution would be associated with a three-fold increase in the number of children with severe deficits (IQ < 80) along with a 5% reduction in the number of children who attain superior function (IQ > 125). In an 11-year follow-up report, Needleman et al. [108] re-examined half of the original population. As compared with those restudied, the ones not restudied had somewhat higher lead levels on earlier analysis, as well as significantly lower IQ scores and poorer teachers' ratings of classroom behavior.

For the subjects reexamined in 1988, impairment in neurobehavioral function was still related to the lead content of teeth shed at the ages of six and seven. The young people with dentine lead levels > 20 ppm had a markedly higher risk of dropping out of high school (adjusted odds ratio, 7.4) and of having a reading disability (odds ratio, 5.8) as compared with those with dentin lead levels < 10 ppm. Higher lead levels in childhood were also significantly associated with lower class standing in high school, increased absenteeism, lower vocabulary and grammatical-reasoning scores, poorer hand-eye coordination, longer reaction times, and slower finger tapping.

In order to avoid the normal array of confounding factors, Bellinger et al. [109] performed a longitudinal analysis of prenatal and postnatal lead exposure and early cognitive development in 249 children. In general, the infants were healthy products of unremarkable pregnancies, with few of the characteristics of infants at increased risk of developmental handicap. Eighty-seven percent of the families were white, and 92 percent were intact. The differences among the families with infants in the three cord-blood lead groups were slight and generally not in the direction expected on the basis of studies of the social correlates of childhood lead exposure. On the basis of lead levels in umbilical-cord blood, children were assigned to one of three prenatal-exposure groups: low (< 3 μg dL^{-1}), medium (6 to 7 μg dL^{-1}), or high ($\geq$ 10 μg dL^{-1}). Development was assessed semiannually, beginning at the age of six months, with use of the Mental Development Index of the Bayley Scales of Infant Development.

Regression methods for longitudinal data were used to evaluate the association between infants' lead levels and their development scores after adjustment for potential confounders. At all ages, infants in the high-prenatal-exposure group scored lower than infants in the other two groups.

McMichael et al. [110] studied the effect of environmental exposure to lead on children's abilities at the age of four years in a cohort of 537 children born during 1979 to 1982 to women living in a community situated near a lead smelter at Port Pirie in Australia. Samples for measuring blood lead levels were obtained from the mothers antenatally, at delivery from the mothers and umbilical cords, and at the ages of 6, 15, and 24 months and then annually from the children. Concurrently, the mothers were interviewed about personal, family, medical, and environmental factors. Maternal intelligence, the home environment, and the children's mental development (as evaluated with use of the McCarthy Scales of Children's Abilities) were formally assessed.

The mean blood lead concentration varied from 9.1 μg dL^{-1} in midpregnancy to a peak of 21.2 μg dL^{-1} at the age of two years. The blood lead concentration at each age, particularly at two and three years, and the integrated postnatal average concentration were inversely related to development at the age of four. Results of multivariate analysis are illustrated in Fig. 5. Within the range of exposure studied, no threshold dose for an effect of lead was evident.

This cohort study indicates that a raised blood lead concentration in early childhood has an independent deleterious effect on mental development as evalu-

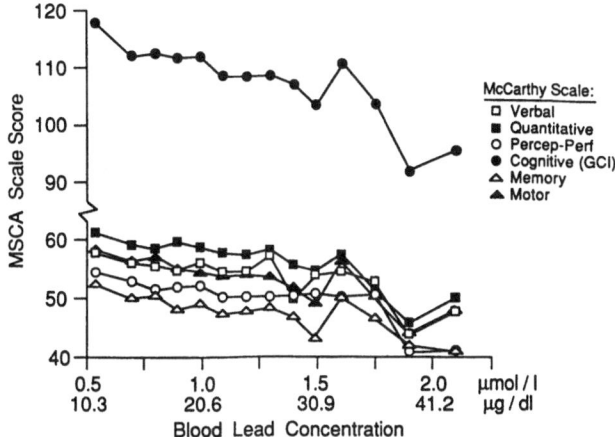

Fig. 5. McCarthy scales of children's abilities (MSCA) scores at the age of 4 vs blood lead concentration at 3 years of age. From: McMichael et al. [110]

ated at the age of four years. This effect was not accounted for by the known and measurable influences of obstetrical, parental, family, and social environmental factors on mental development. The results of this analysis and those of an earlier analysis of the children at the age of two years suggest that increased exposure to lead results in a developmental deficit, not just developmental delay.

Bhattacharya et al. [111] found that abnormalities in children's abilities to maintain physical balance were significantly associated with PbB. Their postural sway on a balance increased by ~ 2.8 cm^2 per μg dL^{-1}. These data suggest that low levels of PbB affect the peripheral nervous system as well as the central nervous system. A sample of their results are illustrated in Fig. 6.

Schwartz and Otto [112] used the large data base available from the Second National Health and Nutrition Examination Survey (NHANES II), conducted between 1976 and 1980 on population samples selected as being representative of the civilian, noninstitutionalized U.S. population. For a subsample of 4,519 youths aged 4–19, there were data available on blood Pb; audiometry and various indicators of neurological development, such as age at which a child first sat up, walked, and spoke. The presence of speech difficulties and hyperactivity were also examined to determine if they were significantly related to lead exposure. The probability of elevated hearing thresholds at 500, 1000, 2000, and 4000 Hz increased significantly ($p < 0.0001$) with increasing PbB (Fig. 7). PbB levels were also significantly related to delays in the age at which children first sat up, walked, and spoke and to the probability that a child was hyperactive. Lead was not related to the probability of a child having a previously diagnosed speech impairment. The results of this large population study are clearly consistent with, and strongly supportive of, the validity of the associations between blood lead and neurobehavioral effects in the smaller populations reviewed earlier.

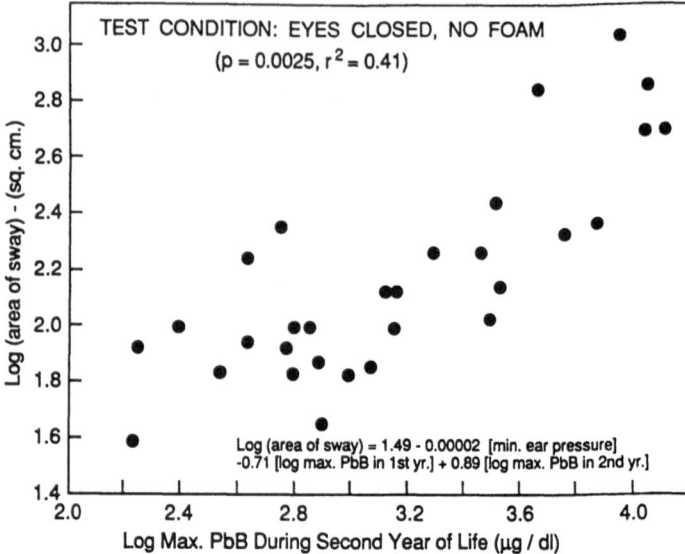

Fig. 6. Log of postural sway of children at 6 yr vs log blood level concentration during second year of life after controlling for blood lead during first year of life. From: Bhattacharya et al. [111]

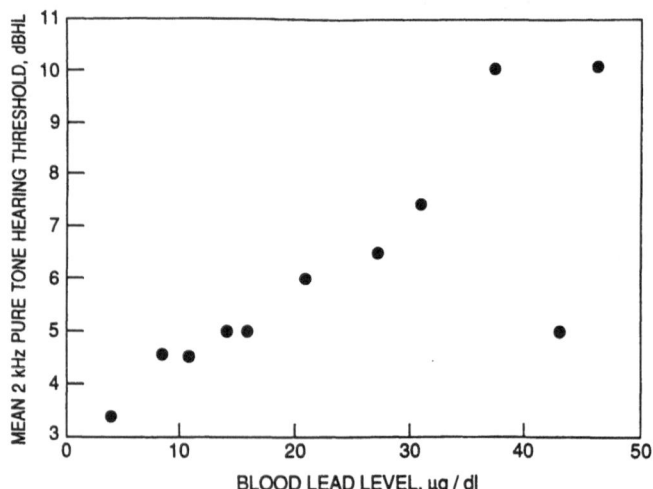

Fig. 7. Relationship of 2 kHz pure tone hearing threshold (right ear) and blood levels in 4519 NHANES II subjects ages 4–19 years. Each point represents the mean hearing threshold of all persons in a 5 μg/dl blood lead range, except for the last point, which represents the mean hearing threshold and mean blood lead for all children with blood lead levels over 35 μg dl^{-1}. From: Schwartz and Otto [112]

In another examination of NHANES II data, Schwartz, Angle, and Pitcher [113] incorporated medical history, physical examination, anthropometric measurements, dietary information (24 h recall and food frequency), laboratory tests, and radiographs in linear regressions of adjusted data from 2,695 children aged 7 years and younger. They reported that 91% of the variance in height, 72% of the variance in weight, and 58% of the variance in chest circumference were explained by six variables: age, race, sex, blood lead level, total calories or protein, and hematocrit or transferrin saturation level.

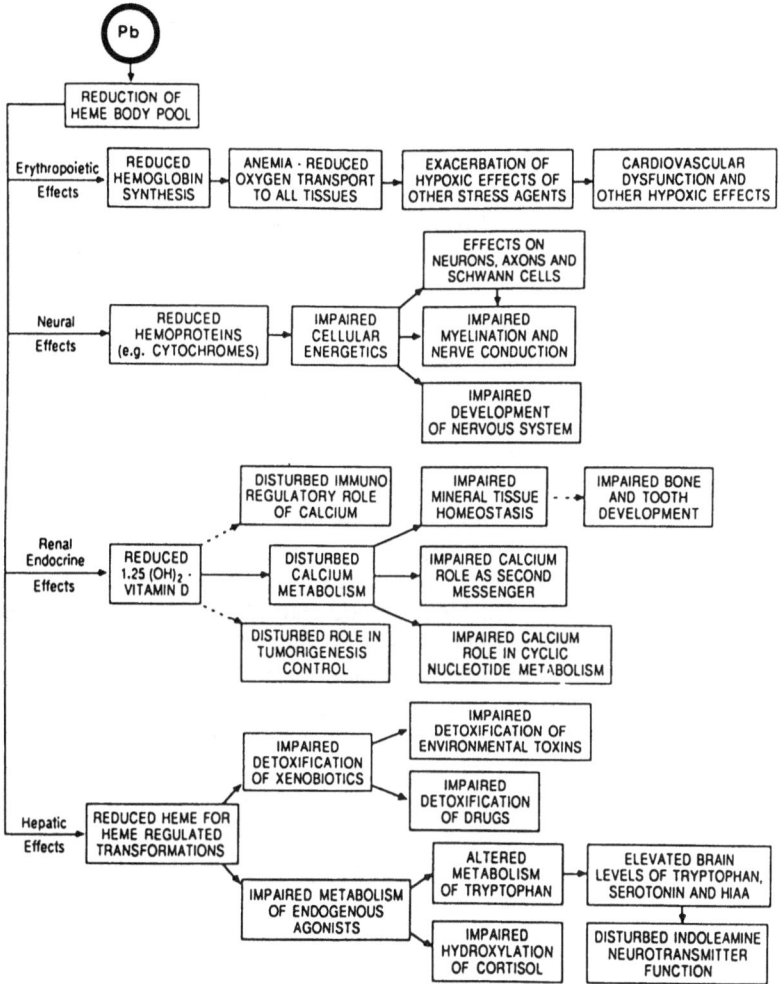

Fig. 8. Multiorgan impact of reductions of heme body pool by lead. Impairment of heme synthesis by lead results in disruption of a wide variety of important physiological processes in many organs and tissues. Particularly well documented are erythropoietic, neural, renal-endocrine, and hepatic effects indicated above by *solid arrows*. Plausible further consequences of heme synthesis interference by lead are indicated by *dashed arrows* (EPA, 1986)

In summary, there are a number of well designed studies which indicate that very low levels of exposure to lead affect neurobehavioral function and development in young children. These various effects appear to be consistent with the effects of lead on heme biosynthesis which have been postulated to lead to erythropoietic, neural, renal endocrine, and hepatic effects in the body, as summarized in Fig. 8 from the 1986 EPA Criteria Document.

Effects of Lead in the Blood on Blood Pressure

The 1986 EPA Criteria Document on Lead [114] also provided a critical review of studies showing associations between blood lead concentrations less than $40\ \mu g\,dL^{-1}$ and blood pressure. It reviewed the influence of a number of environmental and nutritional factors affecting blood pressure in experimental and epidemiological studies. Among environmental factors that have been associated with blood pressures are lead (Pb) and noise. Among dietary factors associated with blood pressure are calcium (Ca), zinc (Zn), phosphorus (P), alcohol consumption, vitamins A and C.

The role of Pb as a pollutant stressor for elevated blood pressure could well be confounded by the well established role of Ca as a suppressor of blood pressure. It is possible that persons with high Ca consumption have both decreased blood pressure and reduced blood Pb due to the competition of both Pb and Ca for the same binding sites. The influence of the other co-factors known to affect blood pressure further complicates the task of establishing the extent to which Pb constitutes a significant risk factor for elevated blood pressure.

A consistent pattern of results emerges from recent investigations of the relations between lower-level lead exposures and increases in blood pressure or hypertension. Khera et al. [115] reported higher blood lead levels in hypertensive patients and those with other cardiovascular diseases than for hospital control subjects. Kromhout and Couland [116] and Kromhout et al. [117] reported associations between hypertension and blood lead among elderly men in the Netherlands. Batuman et al. [118] reported an association between hypertension and chelatable lead burdens in veterans. Moreau et al. [119] reported significant associations ($p < 0.001$) between blood lead levels and a continuous measure of blood pressure among French policemen after controlling for important potential confounding variables such as age, body mass index, smoking, and drinking. Weiss et al. [120] reported that after correction for previous systolic blood pressure, body mass index, age, and smoking, a high level of blood lead was a significant predictor of subsequent elevation of systolic pressure in policemen in Boston. Sharp et al. [121] examined relationships between blood lead concentration and blood pressure in San Francisco bus drivers. The analysis was limited to subjects not on treatment for hypertension (n = 288). The blood lead concentration varied from 2 to $15\ \mu g\,dL^{-1}$. While the findings were not statistically significant, they did suggest effects of lead exposure at lower blood lead concentrations than those previously linked with increases in blood pressure. In a follow-up study, Sharp et al. [122]

examined the relationship between blood pressure and blood lead concentration in 51 bus drivers who were treated for hypertension. These drivers were a subset of a representative sample (n = 342) of the driver population (n = ~ 2000), and were not selected for hypertension or lead exposure. Blood lead concentrations ranged from 2–24 $\mu g\,dL^{-1}$. There were 33 subjects treated primarily with diuretics, and 18 subjects were treated with beta blockers. There was a significant mean difference of 12 mmHg in diastolic BP over the range of observed Pb in blood (2.0 to 11.4 $\mu g\,dL^{-1}$) in subjects treated with beta blockers. Thus, beta blocker therapy may be less effective in reducing diastolic pressure in individuals with elevated PbB, even at PbB levels associated with exposures below the current ambient standard and far below the current occupational standard.

In a large population study, Pocock et al. [123] evaluated the relationships between blood lead concentrations, hypertension, and renal function indicators in a clinical survey of 7735 middle-aged men from 24 British towns. The associations between systolic blood pressure and blood lead levels, though small in magnitude, was statistically significant ($p < 0.01$). Analyses of data for men categorized according to blood level concentrations indicated increases in blood pressure only at lower blood lead levels; no further significant increments in blood pressure were observed at higher blood lead levels.

An ideal opportunity to separate the role of Pb from a wide range of potentially confounding nutritional factors was presented by the large data set from NHANES II, a random stratified sample of the U.S. population. Pirkle et al. [124] described the results of their analyses of the data for 40–59 year old white males from this survey population. After adjustment for age, body mass index, all measured nutritional factors and blood biochemistry factors in a multiple linear regression model, the relationships of both systolic and diastolic blood pressures to blood Pb was statistically significant ($p < 0.01$). Further analyses of NHANES II data for adults aged 20–74 were reviewed by Schwartz [125].

The population mean blood Pb levels dropped by 37% between 1976 and 1980, due to reductions in the amount of Pb used in gasoline. This much reduction in blood Pb in this population would be expected to result in a 17.5% decrease in diastolic blood pressure ≥ 90 mmHg, a level used to define hypertension.

Considering the relatively unusual nature of the blood-Pb/blood-pressure relation (i.e. characterized by large initial increments in blood pressure at relatively low blood Pb levels, followed by leveling off of blood pressure increments at high blood Pb levels), it is not surprising that it was not anticipated by results of studies in animals. Many animal studies emphasize results from exposures at higher dose levels, where results tend to be more definitive. The human results were, however, consistent with biphasic blood pressure increases observed in response to small PbB increases in the rat [126, 127, 128] when rats were treated with low doses of lead. The unusual exposure-response relation may also account for the failure of earlier human studies to find consistent relations between blood pressure and blood Pb in study groups with mild-to-moderate elevations of blood Pb concentrations.

In summary, the use of a very large set of high-quality data covering a wide range of possibly confounding variables allowed a clean-cut determination of the effects of blood Pb on blood pressure for a relatively low range of blood Pb concentrations (5–35 $\mu g\,dL^{-1}$). This association between relatively small elevations of PbB and elevated blood pressure may have significant public health impact because hypertension is a recognized risk factor for cardiovascular disease.

Acknowledgements

The preparation of this paper was supported as part of a Center Program supported by Grant ES-00260 from the National Institute of Environmental Health Sciences and Grant CA 13343 from the National Cancer Institute. Much of the material on O_3, NO_x, SO_x, and CO was taken from a paper prepared for the Congressional Research Service entitled "Health Benefits from Controlling Exposures to Criteria Air Pollutants (January 1989). The material on Pb includes much of the contents of a review paper by the author published in Environmental Research and entitled "Lead and human health: Background and recent findings" (Vol. 51, 1–24, 1990).

References

1. US EPA (1986) Air quality criteria for ozone and other photochemical oxidants, Vol II, EPA/600/8-84/0206F, ECAO, US EPA, Research Triangle Park, NC, August 1986
2. US EPA (1988) Review of the national ambient air quality standards for ozone—Preliminary assessment of scientific and technical information, OAQPS Draft Staff Paper, November 1988
3. OTA (1988) Urban ozone and the Clean Air Act: Problems and proposals for change. Oceans and Environment Program, Office of Technology Assessment, US Congress, Washington, DC 20510-8025 (April 1988)
4. Rombout PJF, Lioy PJ, Goldstein BD (1986) J Air Pollut Control Assoc 36: 913
5. Hazucha MJ (1987) J Appl Physiol 62: 1671
6. McDonnell WF, Horstman DH, Abdul-Salaam S, House DE (1985) Am Rev Respir Dis 131: 36
7. Kagawa J (1984) Int Arch Occup Environ Health 53: 345
8. Shephard RJ, Urch B, Silverman F, Corey PN (1983) Environ Res 31: 125
9. Drechsler-Parks DM, Bedi FJ, Horvath SM (1987) Exp Geront 22: 91
10. Reisenauer CS, Koenig JQ, McManus MS, Smith MS, Kusic G, Pierson WE (1988) J Air Poll Contr Assoc 38: 51
11. Koenig JQ, Covert DS, Marshall SG, van Belle G, Pierson WE (1987) Am Rev Respir Dis 136: 1152
12. Linn WS, Jones MP, Bachmeyer EA, Spier CE, Mazur SF, Avol EL, Hackney JD (1980) Am Rev Respir Dis 121: 243
13. McDonnell WF, Horstman DH, Abdul-Salaam S, Raggio LJ, Green JA (1987) Toxicol Indust Health 3: 507
14. Linn WS, Shamoo DA, Venet TG, Spier CE, Valencia LM, et al. (1983) Arch Environ Health 38: 278
15. Solic JJ, Hazucha MJ, Bromberg PA (1982) Am Rev Respir Dis 125: 664
16. Farrell BP, Kerr HD, Kulle TJ, Sauder LR, Young JL (1979) Am Rev Respir Dis 119: 725
17. Folinsbee LJ, Bedi FJ, Horvath SM (1980) Am Rev Respir Dis 121: 431
18. Hackney JD, Linn WS, Mohler JG, Collier CR (1977) J Appl Physiol 43: 82
19. Horvath SM, Gliner JA, Folinsbee LJ (1981) Am Rev Respir Dis 123: 496

20. Kulle TJ, Sauder LR, Kerr HD, Farrell BP, Bermel MS, Smith DM (1982) Am Ind Hyg Assoc J 43: 832
21. Tepper JS, Costa DL, Lehmann JR, Weber MF, Hatch GE (1989) Amer Rev Respir Dis 140: 493
22. McDonnell WF, Horstman DH, Hazucha MJ, Seal E, Haak ED, Salaam SA, House DE (1983) J Appl Physiol 54: 1345
23. Kulle TJ, Sauder LR, Hebel JR, Chatham MD (1985) Am Rev Respir Dis 132: 36
24. Spektor DM, Lippmann M, Lioy PJ, Thurston GD, Citak K, James DJ, Bock N, Speizer FE, Hayes C (1988) Am Rev Respir Dis 137: 313
25. Kinney PL, Ware JH, Spengler JD, Dockery DW, Speizer FE, Ferris BG (1989) Am Rev Respir Dis 139: 56
26. Lippmann M, Lioy PJ, Leikauf G, Green KB, Baxter D, Morandi M, Pasternack B, Fife D, Speizer FE (1983) Adv in Modern Environ Toxicol 5: 423
27. Spengler JD, Keeler GJ, Koutrakis P, Ryan PB, Raizenne M, Franklin CA (1989) Environ Health Perspect 79: 43
28. Linn WS, Avol EL, Shamoo DA, Peng RC, Valencia LM, Little DE, Hackney JD (1988) Toxicol Indust Health 4: 505
29. Folinsbee LJ, McDonnell WF, Horstman DH (1988) J Air Pollut Contr Assoc 38: 28
30. Horstman D, Folinsbee L, Ives P, Abdul-Salaam S, McDonnell W (1990) Am Rev Respir Dis 142: 1158
31. Hayes SR, Moezzi M, Wallsten TS, Winkler RL (1987) An analysis of symptom and lung function data from human controlled ozone exposure studies. Draft Final Report, San Rafael, CA: Systems Applications, Inc.
32. Linder J, Herren D, Monn C, Wanner HU (1988) Schweiz Ztschr Sportmed 36: 5
33. Avol EL, Linn WS, Shamoo DA, Spier CE, Valencia LM, Venet TG, Trim SC, Hackney JD (1987) J Air Pollut Contr Assoc 37: 158
34. Avol EL, Linn WS, Shamoo DA, Valencia LM, Anzar UT, Hackney JD (1985) Am Rev Respir Dis 132: 619
35. McDonnell WF III, Chapman RS, Leigh MW, Strope GL, Collier AM (1985) Am Rev Respir Dis 132: 875
36. Spektor DM, Lippmann M, Thurston GD, Lioy PJ, Stecko J, O'Connor G, Garshick E, Speizer FE, Hayes C (1988) Am Rev Respir Dis 138: 821
37. Altshuller AP (1977) J Air Pollut Contr Assoc 27: 1125
38. National Research Council (1977) In: Ozone and Other Photochemical Oxidants, pp. 323–87, Washington, DC: NAS
39. Hammer DI, Hasselblad V, Portnoy B, Wehrle PF (1974) Arch Environ Health 28: 255
40. Schwartz J, Zeger S (1990) Am Rev Respir Dis 141: 62
41. Schoettlin CE, Landau E (1961) Publ Health Rep 76: 545
42. Holguin AH, Buffler PA, Contant CF, Jr, Stock TH, Kotchmar D, Hsi BP, Jenkins DE, Gehan BM, Noel LM, Mei M (1985) Trans APCA, TR-4, 262–280
43. Whittemore A, Korn E (1980) Am J Public Health 70: 687
44. Ostro BD, Rothschild S (1989) Environ Res 50: 238
45. Bates DV, Sizto R (1989) Environ Health Perspect 79: 69–72
46. Ozkayanak H, Kinney PL, Burbank B (1990) Preprint 90-150.6 1990 Annual Meeting of Air and Waste Management Assoc, AWMA, Pittsburgh, PA
47. Foster WM, Costa DL, Langenback EG (1987) J Appl Physiol 63: 996
48. Kenoyer JL, Phalen RF, Davis JR (1981) Exp Lung Res 2: 111
49. Schlesinger RB, Driscoll KE (1987) J Toxicol Environ Health 20: 125
50. Lippmann M, Gearhart JM, Schlesinger RB (1987) Appl Ind Hyg 2: 188
51. Driscoll KE, Vollmuth TA, Schlesinger RB (1986) Fund Appl Toxicol 7: 264
52. Kehrl HR, Vincent LM, Kowalsky RJ, Horstman DH, O'Neill JJ, McCartney WH, Bromberg PA (1987) Am Rev Respir Dis 135: 1174
53. Seltzer J, Bigby BG, Stulbarg M, Holtzman MJ, Nadel JA (1986) J Appl Physiol 60: 1321
54. Koren HS, Devlin RB, Graham DE, Mann R, McGee MP, Horstman DE, Kozumbo WJ, Becker S, House DE, McDonnell WF, Bromberg PA (1989a) Am Rev Respir Dis 139: 407
55. Koren H, Devlin RB, Graham D, Mann R, McDonnell WF (1989b) In: Schneider T, Lee SD,

Wolters GJR, Grant LD (eds) Atmospheric ozone research and its policy implications, Elsevier, Amsterdam, p 745

56. Graham DE, Koren HS (1990) Am Rev Respir Dis 142: 152
57. Overton JH, Miller FJ (1987) Modelling ozone absorption in lower respiratory tract. 1987 Annual Meeting of Air Pollut Contr Assoc, June 1987, New York; Preprint 87-99.4
58. Gerrity TR, Wisster MJ (1987) Experimental measurements of the uptake of ozone in rats and human subjects. 1987 Annual Meeting of Air Pollut Contr Assoc, June 1987, New York; Preprint 87-99.3
59. Amoruso MA, Goldstein BD (1988) Toxicologist 8: 197
60. Huang Y, Chang LY, Miller FJ, Graham JA, Ospital JJ, Crapo JD (1988) Am J Aerosol Med 1: 180
61. Tyler WS, Tyler NK, Last JA, Gillespie MJ, Barstow TJ (1988) Toxicology 50: 131
62. Detels R, Tashkin DP, Sayre JW, Rokaw SN, Coulson AH, Massey FJ, Wegman DH (1987) Chest 92: 594
63. Knudson RJ, Lebowitz MD, Helerg CJ, Burrows B (1983) Am Rev Respir Dis 127: 725
64. US EPA-ECAO Air quality criteria for ozone and other photochemical oxidants. EPA/600/8-84/0206F. Environmental Criteria and Assessment Office, Research Triangle Park, NC 27711
65. State of California Air Resources Board Effects of ozone on health – technical support document. Air Resources Board, Sacramento, CA 95812
66. Kilburn KH, Warshaw R, Thornton JC (1985) Am J Med 79: 23
67. Schwartz J (1989) Environ Res 50: 309
68. Lioy PJ, Vollmuth TA, Lippmann M (1985) J Air Pollut Contr Assoc 35: 1068
69. Raizenne ME, Burnett RT, Stern B, Franklin CA, Spengler JD (1989) Environ Health Perspect 79: 179
70. Stacy RV, Seal E Jr., House DE, Green J, Roger LJ, Raggio L (1983) Arch Environ Health 38: 104
71. Holtzman MJ, Cunningham JH, Sheller JR, Irsigler GB, Nadel JA, Boushey H (1979) Am Rev Respir Dis 120: 1059
72. Utell MJ, Morrow PE, Speers DM, Darling J, Hyde RW (1983) Am Rev Respir Dis 128: 444
73. Last JA (1989) Environ Health Perspect 79: 115
74. Kleinman MT, Phalen RF, Mautz WJ, Mannix RC, McClure TR, Crocker TT (1989) Environ Health Perspect 79: 137
75. Graham JA, Gardner DE, Blommer EJ, House DE, Menache MG, Miller FJ (1987) J Toxicol Environ Health 21: 113
76. Pinkerton KE, Brody AR, Miller FJ, Crapo JD (1989) Am Rev Respir Dis 140: 1075
77. US EPA Review of the national ambient air quality standards for nitrogen oxides: Assessment of scientific and technical information (1982) OAQPS staff paper. EPA-450/5-82-002. Office of Air Quality Planning and Standards, Research Triangle Park, NC (Aug. 1982)
78. Bauer MA, Utell MJ, Morrow PE, Speers DM, Gibb FR (1966) Am Rev Respir Dis 134: 1203
79. Roger LJ, Horstman DH, McDonnell WF, Kehrl H, Seal E, Chapman RS, Massaro EJ (1985) Toxicologist 5: 70
80. Mohsenin V (1987) J Toxicol Environ Health 22: 371
81. Linn WS, Shamoo DA, Avol EL, Whynot JD, Anderson KR, Venet TG, Hackney JD (1986) Arch Environ Health 41: 292
82. Mohsenin V (1987) Am Rev Respir Dis 136: 1408
83. Koenig JQ, Covert DS, Pierson WE, McManus MS (1988) Am Rev Respir Dis 137/c (4): A169
84. Smeglin AM, Roberts NJ Jr, Morrow PE, Utell MJ (1986) Am Rev Respir Dis 133/c (4): A216
85. Frampton MW, Smeglin AM, Roberts NJ Jr, Finkelstein JN, Morrow PE, Utell MJ (1987) Am Rev Respir Dis 135(4): A58
86. Kulle TJ, Goings SAJ, Sauder LR, Green DJ, Clements ML (1987) Am Rev Respir Dis 135(4): A58
87. Miller FJ, Graham JA, Raub JA, Illing JW, Menache MG, House DE, Gardner DE (1987) J Toxicol Environ Health 21: 99
88. Lioy PJ, Waldman JM (1989) Environ Health Perspect 79: 15
89. Koenig JQ, Pierson WE, Horike M (1983) Am Rev Respir Dis 128: 221
90. Koenig JQ, Covert DS, Pierson WE (1989) Environ Health Perspect 79: 173
91. Bauer MA, Utell MJ, Speers DM, Gibb FR, Morrow PE (1988) Am Rev Respir Dis 137(4): A167

92. Schlesinger RB (1989) Environ Health Perspect 79: 121
93. Spektor DM, Yen BM, Lippmann M (1989) Environ Health Perspect 79: 167
94. Amdur MO, Chen LC (1989) Environ Health Perspect 79: 147
95. Lippmann M, Gearhart JM, Schlesinger RB (1987) Appl Ind Hyg 2: 188
96. Schlesinger RB (1986) In: Lee SD, Schneider T, Grant LD, Verkerk PJ (eds) Aerosols. Lewis, Chelsea, MI, p 617
97. Speizer FE (1989) Environ Health Perspect 79: 61
98. Wichmann HE, Mueller W, Allhoff P, Beckmann M, Bochter N, Csicsaky MJ, Jung M, Molik B, Schoeneberg G (1989) Environ Health Perspect 79: 89
99. Dassen W, Brunekreef B, Hoek G, Hofschreuder P, Staatsen B, deGroot H, Schouten E, Biersteker K (1986) J Air Pollut Contr Assoc 36: 1223
100. Allred EN, Bleecker ER, Chaitman BR, Dahms TE, Gottlieb SO, Hackney JD, Pagano M, Selvester RH, Walden SM, Warren J (1989) N Engl J Med, 321: 1426
101. ATSDR (Agency for Toxic Substances and Disease Registry) (1988). The nature and extent of lead poisoning in children in the United States. USDHHS, Public Health Service, Atlanta, GA, July 1988
102. Yaffe Y, Flessel CP, Wesolowski JJ, del Rosario A, Guirguis GN, Matias V, Degarmo TE, Coleman GC, Gramlich JW, Kelly WR (1983) Arch Environ Health 38: 227
103. Bellinger D, Needleman MC, Bromfield R, Nimtz M (1984) Biol Trace Elements Res 6: 207
104. Winneke G, Kramer U, Brockhaus A, Ewers U, Kujanek G, Lechner H, Janke W (1983) Int Arch Occup Environ Health 51: 231
105. Needleman HL, Gunnoe C, Leviton A, Reed R, Peresie H, Maher C, Barrett P (1979) N Engl J Med 300: 689
106. Needleman HL (1984) Comments on chapter 12 and appendix 12C, Air Quality Criteria for Lead. US Environmental Protection Agency, Central Docket Section, Docket ECAO-CD-81-2 IIA.E.C.1.20. Washington, DC
107. Needleman HL, Leviton A, Bellinger D (1982) N Engl J Med 306: 367
108. Needleman HL, Schell A, Bellinger D, Leviton A, Allred EN (1990) N Engl J Med 322: 83
109. Bellinger D, Levitan A, Waternaux C, Needleman H, Rabinowitz M (1987) N Engl J Med 316: 1037
110. McMichael AJ, Baghurst PA, Wigg NR, Vimpani GV, Robertson EF, Roberts RJ (1988) N Engl J Med 319: 468
111. Bhattacharya A, Shukla R, Bornschein R, Dietrich K, Kopke JE (1988) Neurotoxicol 9: 327
112. Schwartz J, Otto D (1987) Arch Environ Health 42: 153
113. Schwartz J, Angle C, Pitcher H (1986) Pediatr 77: 281
114. EPA (1986) Air Quality Criteria for Lead. EPA-600/8-83-028 U.S. EPA, Environmental Criteria and Assessment Office, Research Triangle Park, NC, June 1986
115. Khera AK, Wibberley DG, Edwards KW, Waldron HA (1980) Int J Environ Stud 14: 309
116. Kromhout D, Couland CL (1984) Eur Heart J 5 (abstr suppl 1): 101
117. Kromhout D, Wibowo AE, Herber FM, Dalderup LM, Heerdink H, Coulander C de L, Zielhuis RL (1985) Am J Epidemiol 122: 378
118. Batuman V, Landy E, Maesaka JK, Weeden RP (1983) N Engl J Med 309: 17
119. Moreau T, Orsaaud G, Juguet B, Busquet G (1982) Rev Epidemol Sante Publique 30: 395
120. Weiss ST, Munoz A, Stein A, Sparrow D, Speizer FE (1986) Am J Epid 122: 800
121. Sharp DS, Osterloh J, Becker CE, Bernard B, Smith AH, Fisher JM, Syme SL, Holman BL, Johnston T (1988) Environ Health Perspect 78: 131
122. Sharp DS, Smith AH, Holman BL, Fisher JM, Osterloh J, Becker CE (1989) Arch Environ Health 44: 18
123. Pocock SJ, Shaper AG, Ashby D, Delves T, Whitehead TP (1984) Br Med J 289: 872
124. Pirkle JL, Schwartz J, Landis JR, Harlan WR (1985) Am J Epidemiol 121: 246
125. Schwartz J (1988) Environ Health Perspect 78: 15
126. Victery W, Vander AJ, Markel H, Katzman L, Shulak JM, Germain C (1982) Proc Soc Exp Biol Med 170: 63
127. Victery W, Vander AJ, Shulak JM, Schoeps P, Julius S (1982) J Lab Clin Med 99: 354
128. Perry Jr HM, Erlanger MW, Perry EF (1988) Environ Health Persp 78: 107

Mortality and Air Pollution

Frederick W. Lipfert

Brookhaven National Laboratory, Upton, NY USA 11973

Summary

Studies linking air pollution and human mortality are reviewed and data are reanalyzed to provide a consistent basis for synthesis. The studies are reviewed in three groups: short duration episodes of high air concentrations; time-series analyses of daily mortality; and cross-sectional analyses of long-term mortality rates. Both sulfur dioxide (SO_2) and particulate matter are implicated in short-term effects on mortality; although most of the studies reviewed are consistent, it is not possible to define thresholds of "no effect". The cross-sectional studies are less consistent, but together suggest that air pollution also influences long-term mortality rates. It is not possible to specify with certainty the pollutants responsible nor to determine whether the long-term effects are additional to the annual sum of short-term effects on mortality.

Introduction

Protection of public health has long been an important force for the control of air pollution, beginning with the convincing evidence of dramatic increases in daily deaths during the episodes of severe air pollution between 1930 and the early 1960s. Although the adverse effects of such acute episodes are well-recognized, debate continues on the effects on human health of chronic exposure to lower levels of air pollution. More than 30 years ago, it was observed that sulfur dioxide (SO_2) seldom exceeded 2 ppm even in polluted (American) cities [1]; today levels are an order of magnitude lower and there is debate whether 2 ppm is acceptable even for occupational exposures. Similarly, at that time, particle concentrations were measured in *milli*grams per cubic meter, whereas current standards use units of micrograms per cubic meter (abbreviated here as $\mu g/m^3$)

Although clean air is widely perceived as an important public good, scientists, economists, and policy makers continue to debate the adequacy of ambient air quality standards. When public and private funds for the control of pollution are limited, the direction of those funds to the most cost-effective ends is an important, if difficult, responsibility. Unfortunately, even after years of regulation, the measures of the benefits purchased by the billions of dollars spent on air pollution control are often elusive.

Measurements of Air Pollution

Early measurements of air pollution in cities focused on two pollutants: sulfur dioxide (SO_2) and particulate matter. Characterization of the latter has progressed from simple dustfall measurements (gravimetric catch in a glass jar), to smokeshade (degree of staining of a filter paper), to total suspended particles (TSP) (gravimetric catch of particles > 50 μm diameter on a glass-fiber filter), to size-classified particle mass. The rationale behind these changes was concern that smaller particles are potentially more injurious to health, since they can penetrate deeper into the lung. In addition, the technology of particle sampling has been influenced by the need for subsequent chemical analysis, for example, to determine the content of sulfates and other ions or trace elements.

SO_2 measurements have progressed from sulfation plates or candles (sulfur deposition gauges), to simple wet-chemical bubblers and their automated counterparts, to automated gas-phase detectors that telemeter results to a central computer. For both SO_2 and particle measurements, averaging times for routine monitoring have progressed from monthly to hourly, although 24-hour samples normally are taken for particle mass.

Present-day air quality control recognizes several additional pollutants as potentially important to health, including oxidants and ozone (O_3), carbon monoxide (CO), nitrogen oxides (NO_2 or NO_x), lead (Pb), organics such as formaldehyde, and various aerosol constituents. Measurement technologies have improved with time, especially for ozone and NO_2. The early (pre-1970) bubbler measurements of NO_2 are particularly suspect in the United States.

Unfortunately, the spatial coverage of urban air monitoring has decreased over the years. In many cities, the extensive networks formerly present have been reduced, so that population exposures are defined inadequately. In addition, much of the air monitoring in the United States is devoted to checking compliance with standards around large point sources, regardless of the impact on populations. Since urban concentrations were higher in the early years (before 1970), epidemiological studies of this period are often of special interest.

A recent development of importance to health studies is the role of indoor air quality. Since populations in temperate climates spend most of their time indoors, their total exposures to air pollution can be dominated by indoor air quality. While some pollutants have important indoor sources (such as NO_2, formaldehyde, and small particles), indoor concentrations of others may be greatly attenuated from outdoor levels (notably SO_2).

Determination of the Effects of Air Pollution

There are several ways to determine the adverse effects of air pollution, or, conversely, the benefits of its control:

1. *Theoretical estimates.* On the basis of physical principles, it is possible to calculate the loss in atmospheric visibility or in sunlight received at the earth's surface due to aerosols, for example. Similarly, the effects of atmospheric CO_2 on climate have been estimated theoretically. However, because living organisms can adapt, generally it is not feasible to make predictive theoretical calculations of the health effects of air pollutants, although it is possible to estimate respiratory deposition, for example, as a function of particle size.

2. *Controlled experiments.* Controlled laboratory and, to a lesser extent, field experiments have been used to determine damage to materials and vegetation. Clinical tests on animals and humans fall into this category, as do bioassays which are often used to assess carcinogenicity. Unfortunately, all these methods suffer from lack of realism. It is difficult to extrapolate from animals to humans, and very high concentrations are usually required to induce measurable effects within a reasonable exposure time. In addition, it is difficult to reproduce the mixtures that make up a realistic polluted urban atmosphere, and the experimental apparatus itself may influence the outcome of the experiment. It is important to note that while the control of air pollutants is largely directed toward one pollutant species at a time, their effects almost always occur in combination. Policy makers have found it convenient to separate mobile and stationary source categories, but atmospheric chemistry does not recognize this distinction nor do the recipients of the resulting complex mixtures.

3. *Uncontrolled "experiments".* Several types of events fall into this final and largest category for the determination of effects, including air pollution episodes (e.g., Donora, London, and New York). Somewhat more difficult to evaluate are the case studies of pollution abatement campaigns (e.g., smoke in London, SO_2 in New York City, and oxidants in Southern California). Case-control studies of accidental or occupational exposures fall into this group. Perhaps the most difficult of all to assess are the so-called "natural experiments", in which different geographic areas have been subjected to different (and changing) mixes of air pollution for many years. Descriptions of this phenomenon involve largely non-experimental data with a finite set of observations; inferences drawn from such a data set may correspond to those described by Leamer [2] as "metastatistics", in which ". . . the researcher's motives and opinions influence his choice of model and his choice of data". For example, the same activities that produce pollution also produce a host of other, potentially confounding, effects, both socioeconomic and physiological, on the residents of the affected communities. It is also important to recognize that many factors unrelated to pollution play a role in these geographical differentials in mortality and morbidity.

Studies of the effects of air pollution often differ from classical epidemiology, which usually attempts to explain the patterns of a particular disease, with environmental agents as a possible causal factor. In contrast, many studies of air pollution focused on the agent(s) and searched for any of several types of health effects that could conceivably be associated with certain physiological responses. However, many of these studies, which be classified as "observational" rather than experimental, employed the traditional epidemiological methods of examining disease trends according to place and time.

Mortality is a useful end point in studying the health effects of air pollution only because of the ready availability of the data. Epidemiological studies undoubtedly would be more definitive if an end point were identified that was unique to air pollution instead of one that is inevitable. The problem of defining premature mortality ("excess deaths") is difficult, and that of assigning causality is even more so. No cause of death is uniquely associated with exposure to air pollution, and those studies that reported widespread effects in large populations from air pollution usually found such effects in the broader cause-of-death categories such as heart disease. (Bronchitis mortality in Britain is an exception to this generality.) This lack of specificity may have resulted in part from undue reliance on the underlying (immediate) cause of death rather than on contributing causes. Also, heart disease is in some sense a "catch-all" cause of death, since it is now such a large fraction of total deaths (currently, about half of all U.S. deaths, if strokes are included).

The determination of air pollution effects on health and mortality continues to challenge both physicians and statisticians. Direct experiments on humans involving long-term clinical exposures or irreversible effects are unethical, so that deductions must be based on indirect observational studies. Because of the many confounding and collinear variables often involved, these studies have become the domain of the statistician, often using multiple regression analyses. However, some important statistical issues have only recently been recognized [3], in spite of almost 20 years of publications on the topic. In addition, statistical association alone can never establish causality, and it is important to invoke physiological considerations when considering the reasonableness of hypotheses generated from the statistical findings.

Considerations of Causality

In considering the health effects of air pollution, certain end points can effectively be studied only by studying groups rather than individuals. Such studies have been called "ecologic". This situation arises when one looks for a small effect (1–5%) to which a large number of people may be exposed (ca. 10^8); thus it is necessary to consider populations too large for personal monitoring in order to achieve statistical significance. However, there are some useful "numerical experiments" that can be performed with ecologic data sets to test causal hypotheses or their converse, circumstantial associations.

1. *Environmental exposures*. The statistical consequence of poorly characterized but unbiased environmental exposures is to bias the regression coefficients and their significance downward, making a true effect harder to detect. Thus, improving the characterization of exposure should increase the coefficients of "true" responses. Such improvements are made by considering the data sets having better monitoring coverage, using mathematical models to smooth between monitoring stations, or by increasing the averaging times for pollutant concentrations. Note that for a pollutant with large hourly variations (as for most "local" pollutants, due to wind movements), the spatial representativeness of the hourly maximum will be worse than for the 24-hour average. Thus the health effects response would have to be highly nonlinear (for all respondents) in order for the hourly maximum concentration to provide a better measure of the population-averaged dose than the 24-hour average.

Biases in exposures can occur when data from source-oriented monitoring stations are used to characterize entire communities (which can be a problem for SO_2 and CO, but less so for NO_2, SO_4^{2-} aerosol, and ozone). Additional biases can occur when the effects of smoking, indoor air or occupational exposures are neglected. These effects can often be discerned by considering specific subsets of the population, for example by age or gender. However, smoking can increase indoor particulate levels substantially, so that the outdoor measurements represent only part of the actual esposure (even to nonsmokers) which reduces the potential differences in exposure by age or gender. A less obvious bias can occur when the monitoring coverage is better (or worse) for a segment of the population that is more likely to be susceptible to the effect.

2. *Lag effects*. A common device in studying time-series phenomena is the analysis of lag effects. It is unreasonable to expect an instant response to a moderate air-pollution insult, so that some degree of (positive) lag is usually found (i.e., response occurs some time after exposure). A finding of statistically significant *negative* lag suggests a spurious association. As the period of the (positive) lag is increased, the cumulative effect should also increase, since different portions of the population may respond differently and an extended cumulative period will encompass more responders. When the lag period begins to exceed the actual response time, the significance of a true effect should begin to decrease. These findings should be robust for a given pollutant and end point across different subsets of the population, assuming that the exposures are equally well-characterized. As characterization of the exposures improves, the effects of lags should become better defined for a "true" effect.

It is also important to account for serial correlation in the above types of analyses. "Local" air pollutants like CO and SO_2 (and perhaps TSP) are expected to exhibit less daily autocorrelation than regional pollutants like SO_4^{2-} aerosol or ozone.

3. *Specificity of responses*. Considering specific diagnoses often involves a trade-off between the accuracy of diagnosis and the introduction of sampling (Poissonian) variance as the numbers of relevant cases become smaller. As non-pollution-related diagnoses are eliminated from the data set, we would expect the regression coefficient to remain constant in absolute terms (cases/dose), but its statistical significance to increase. Consideration of the non-significance of non-pollution-related diagnoses must be keyed to the magnitude of the sampling variances, to insure that non-significance is not simply due to the small numbers of cases.

There are also some important ecologic issues: for all-cause mortality due to a regionally distributed pollutant, there is no issue of ecologic exposure since all people are at risk and are also exposed. However, this may not be true in other cases; for example, a time-series analysis of terminally ill patients. Since they are likely to be hospitalized, they are unlikely to be exposed to most outdoor air pollutants. Conversely, this will not be a problem with studies of hospital *admissions*.

4. *Collinearity issues*. Collinearity is often an issue in ecologic regression analysis, and must be considered on two levels: collinearity among pollutant species, and collinearity between pollutant and other explanatory variables such as weather or socioeconomic lifestyles. In the first instance, we must recognize that pollutants do not occur in the natural atmosphere one at a time and that different people may be sensitive to different species. Also, species may interact, such as the adsorption of

gases onto particles. Thus, it may be impossible to separate the effects of different species.

Collinearity between pollutants and other variables is a more serious matter with respect to causality. If pollutant x is highly correlated with non-pollutant variable y, we must look to exogenous evidence to support our decision on which variable to select; it does not automatically follow that such a situation precludes finding a causal relationship for the pollutant in question. For example, sulfate aerosol tends to be correlated with the age of housing stock (since in the United States sulfur oxide concentrations tend to be higher in the parts of the country that were settled first, have older communities, and often use the locally-available fuels that gave rise to early industrial development). Since living in an old house would not generally be considered a health risk factor, there would be no reason *a priori* to include this variable in a regression model. Similar considerations may apply to certain weather variables, but, in this instance, the hypothesis could be tested by performing time-series analyses in regions where the pollutant-weather relationships differ.

In general, one encounters "shades of gray" on collinearity rather than black-white issues. The general case involves selecting from alternative regression models; as the numbers of independent variables are increased, the differences in overall fit tend to become small. It is important to realize that the statistics used to compare models and coefficients themselves have distributions, and that undue importance should not be attached to a given realization (i.e., data set) or to achieving arbitrary levels of statistical significance. However, statistical tests should be used to decide whether the alternative models are really different; such tests often involve comparing the distributions of residuals.

Content of This Chapter

Three types of epidemiological studies have been used to explore relationships between air pollution and premature mortality: studies of air pollution episodes as unique events, time-series analysis of daily variations in mortality, and analysis of geographic variations in long-term (annual) death rates. Each type of analysis has its own strength and weaknesses, but most observers would probably agree that, together, they indicate a (positive) association between mortality rates and air pollution. However, questions remain as to the meaning of some of the relationships, especially in the context of application to risk assessment and cost-benefit analysis.

This chapter will explore various interpretations of the many studies in the literature (an exploration that must, in part, be subjective), reanalyze some of the data, present some new results dealing with the effects of chronic exposures in the United States, and suggest avenues for future research. The review primarily encompasses works published in the peer-reviewed literature, but also unpublished reports from various sources (the "gray" literature).

There have been many previous reviews of studies of mortality and air pollution [4–8], but an understanding of some of the older studies is often required to evaluate the newer ones. Although advanced statistical analysis techniques are often necessary for such studies, I have minimized statistical jargon and notation in reviewing the material.

Short-Term (Acute) Effects on Mortality

Studies of acute responses to air pollution (fatalities that occur within a few days of the exposure) can be divided into two groups: episodes, during which air quality deteriorated so markedly that the "event" during which a mortality response might be expected is clearly delineated; and relatively longer periods of routine monitoring, which may or may not include episodes, which must be studied using the methods of time-series analysis.

General Considerations

Temporal variations in mortality and morbidity have often been used to deduce the effects of air pollution, beginning with the notorious episodes of many years ago. In a severe episode, the effects may be easy to identify, by comparing the rates of mortality or sickness before, during, and after the episode. Since episodes are often caused by adverse meteorological conditions (often cold and foggy) which increase concentrations of *all* air pollutants, it may not be obvious which specific pollutant is responsible for the observed effects, or whether the weather *per se* played a role; the analysis of short-term episodes is less likely to be confounded by longer term cyclical phenomena such as the normal seasonal cycle, influenza epidemics, and holidays. In general, bivariate correlations will not suffice to decide whether a potentially confounding variable should be included in the analysis, since the covariance structure may change in ways that are not apparent when variables are considered singly.

Fortunately, severe episodes are now very infrequent in most of the developed world (the period of recurrence now seems to be about 10 years or so); as a result, time-series analysis for deducing the health effects of air pollution requires more sophisticated techniques to detect the smaller signals over and above the statistical "noise". Such analyses are also required to deduce levels of air pollution appropriate for ambient air-quality standards.

Time-series analyses for longer periods, of a year or more, must also deal with co-pollutants having similar temporal patterns and with the confounding factors of other temporal influences in mortality rates, which can include epidemics, holiday and weekend effects (or reporting artifacts), seasonal cycles, and long-term changes in the population. As an example of possible confounding, Fig. 1 shows daily mortality and air pollution in London, averaged by day of the week for fourteen winters [9]. Sundays tend to be low in both pollution (because of reduced commercial activity) and mortality, but part of the mortality decrement is regained on Monday, strongly suggesting a reporting artifact. If Sundays and Mondays are averaged, the slope of mortality versus pollution becomes almost constant and a

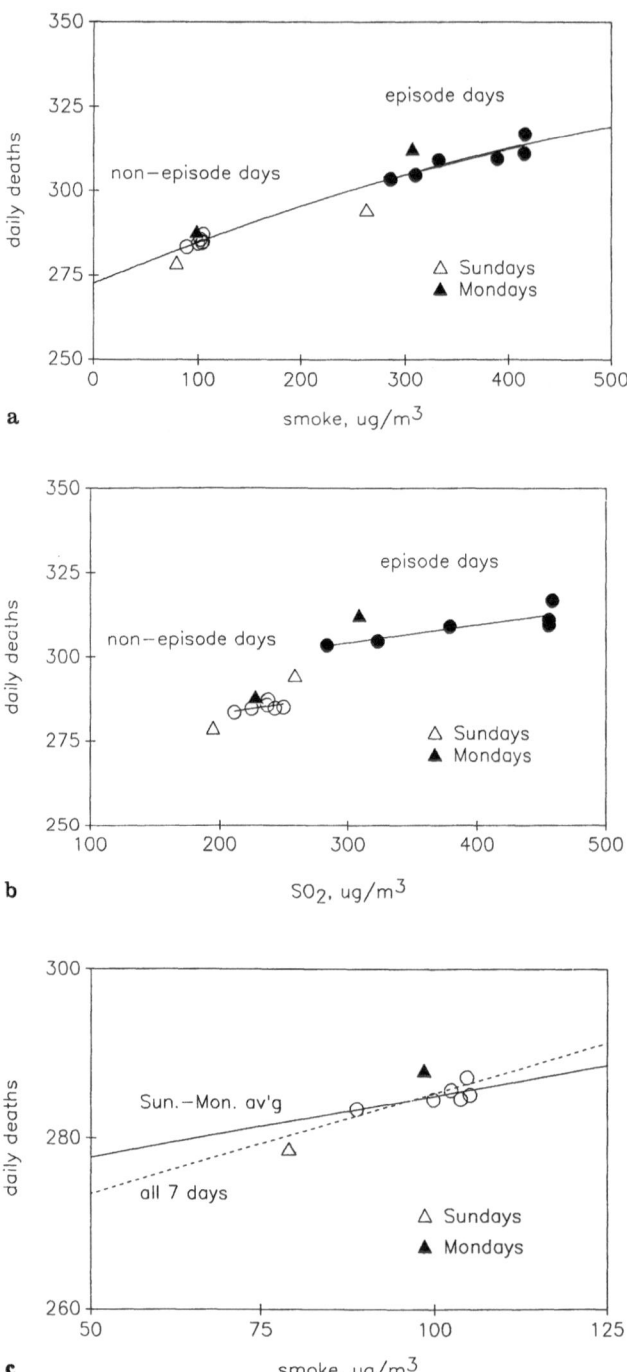

Fig. 1a–c. London mortality, averaged by day of week. **a** By average smoke concentration. **b** By average SO_2 concentration. **c** Non-episode days, by average smoke concentration. Data from Mazumdar et al. (1980) [9]

better fit is obtained. Thus, it is important to account for the day of the week (and holidays) in trying to deduce the true effects on mortality due to air pollution. Various types of moving averages or "filtering" have been used to deal with cyclical mortality variations unrelated to pollution, and with serial correlation.

In summary, a valid long-term time-series analysis should include all pollutants that are physiologically plausible, all other cyclical variables with plausible effects on mortality, and consideration of any long-term trends. Long-term trends could include abatement of air pollution, improvement of medical care and living conditions (including heating and air conditioning), and changes in the resident population (due to migration, for example), as well as changes in the methods of measuring pollutant levels or the health end point. These general requirements may also apply to the study of individual episodes, depending on the length of the time periods, either during the episode or for the "control" periods.

Studies of Episodes

The common factors associated with the well-known air pollution episodes of 1930–60 included stagnant meteorology, fog, and cool or cold weather, which resulted in high levels of pollutant emissions from space heating as well as reduced dispersion. These conditions persisted for several days, causing both people and animals to sicken and die at rates far above those expected for that time of the year.

The Meuse Valley, Belgium, December 1930 [10]

The Meuse Valley episode occurred along a 20-km stretch of the Meuse River between Liege and Huy, which contained 27 factories producing steel, glass, and zinc [11]. Very little is known about this episode (even the population at risk was uncertain [1]), but over 60 "excess" deaths were reported, about ten times the estimated normal number. Since there was no air monitoring in operation, pollution levels could only be estimated from dispersion calculations, which gave an estimate of 10 ppm for SO_2 [10]. Estimates for particulates were not made, but if there was the same proportionality with SO_2 that was observed in other locations, particulate levels would have been in the range 10–20 mg/m³. Fluoride was also implicated in this episode [11].

Donora, PA, October 1948 [12, 13]

Donora, an industrial location near Pittsburgh, was afflicted by a heavy, stagnating smog for four days in October 1948. Fifteen people died on the third day, two on the fourth day, and one more on the twelfth day (two additional deaths were later attributed to the episode). The expected death rate was about one in two days. Again, no air monitoring was in place at the time, and estimates were based on subsequent measurements: 0.6 ppm SO_2, 4.5 mg/m³ total suspended particulates (TSP), including high levels of metallic particles such as cadmium, zinc, and lead. A follow-up study ten years later [14] indicated higher mortality rates for persons exposed to the episode and affected by it, but could not distinguish whether these individuals had preexisting conditions that might also have played a role in their survival rates. Mills [13] presents a detailed personal account of this episode and laid blame on sulfuric acid, a by-product of zinc production.

London, England, December 1952 et seq. [15]

London had been notorious for its "pea soup" fogs for many years and suffered deadly air-pollution episodes on several occasions before and after the major episode that occurred in December 1952 [1]. The statistics of seven such episodes are reviewed by Brasser et al. [16]; the numbers of excess deaths ranged from 200 to 3900. Air quality was determined by the peroxide method for SO_2 and by the darkness of filter stains for particulates ("British smoke" or just "smoke" when referring to British data). The latter measurement is not gravimetric, although the results are reported in units of $\mu g/m^3$. A comparison with other particulate sampling methods is given by Ball and Hume [17].

The December 1952 episode affected all age groups, especially those over age 45. Both central London and the outlying districts (as far away as 50 km [15]) recorded sharp increases in weekly mortality; the largest increases were for bronchitis, pneumonia, and respiratory tuberculosis, although large increases were also noted for deaths due to coronary heart disease and myocardial degeneration. Levels of air pollution reached about 4 mg/m^3 for both SO_2 and smoke (48-hour averages), although there is some question as to the maximum smoke level because of saturation of the filters and the use of visual colorimetric evaluation [5]. Since pollution levels at the height of the episode were averaged over 48 hours, the peaks may have been considerably higher [18]. The principal source of air pollution was the burning of soft coal for home heating in inefficient open grates; as a result, both residential and commercial portions of the city were affected.

Although London air quality improved by an order of magnitude through the application of smoke controls during the 1960s, relatively high levels of pollution can still be approached under adverse meteorological conditions. In December 1975, smoke reached about 550 $\mu g/m^3$ and SO_2 about 1000 $\mu g/m^3$ (24-hour averages); excess mortality for the week was of the order of 6–11% [19].

New York City, November 1953

Like London, New York had several episodes in the 1950s and 1960s. These were analyzed by several authors [20, 21] and some similarities to the London experiences were noted (the absence of heavy fog being a notable exception). During the episode of November 17–21, SO_2 reached 0.86 ppm and smokeshade, 8.5 COHs [21]. Winds were calm up to 5000 ft. altitude at the height of the episode. Eye irritation and respiratory complaints were received from all parts of the city, indicating that a large area was affected. Mortality for all ages above 45 was affected. Compared to London, in New York, bronchitis was not a major factor, the percentage of excess deaths was lower, and SO_2 levels were somewhat lower while particulates were much lower (reflecting the generally cleaner fuels used in New York). The SO_2 measuring system used in New York City at that time is believed to be comparable with that of London, but the tape sampler used for "smoke" was slightly different and data are reported in units of "coefficient of haze" (COH). (For the quantitative analysis and data plots below, COH units were converted to equivalent gravimetric units as given in [4] [approximately 0.1 mg/m³ per COH unit]. This conversion factor agrees best with actual calibration data at the higher concentration levels [5, 19]).

Paris, December 1972–January 1973

Loewenstein and colleagues [22] describe two episodes of excess mortality in Paris: December 1979, with 45% excess mortality for the month, attributed mainly to an epidemic of influenza, and December 25, 1972–January 14, 1973, with 22% excess mortality reported for the period of 20 days. Pollution was near the normal monthly average in the first case, and reached 0.26 mg/m³ for black smoke and 0.66 mg/m³ for "acidité forté" (a measure of SO_2) for the second case. Lowenstein's analysis does not clearly separate the various contributing factors: influenza, a holiday period, cold temperatures, and air pollution. If the SO_2 measure was comparable to those for London and New York, one would conclude that the observed excess mortality in Paris was higher than expected. Part of this may have resulted from including "excess" mortality for an entire month after the episode.

Pittsburgh, PA, November 1975 [23]

This event is classified as an "episode" mainly because industrial curtailment was ordered by the U.S. Environmental Protection Agency (EPA) in response to deteriorating air quality. SO_2 reached 0.2 mg/m³ (0.08 ppm) and TSP, 0.77 mg/m³. EPA estimated that there were 14 excess deaths for this event [23]. Changes in pulmonary function for a subset of sensitive school children were also reported [24]. A more extensive study of acute health effects in Pittsburgh is discussed below with the time-series analyses.

Central and Western Europe, January 1985 [25]

The most recent air pollution episode may have affected several countries, including Eastern Europe, but health effects have been analyzed in only a few. In West Germany, SO_2 levels reached 0.8 mg/m³ and suspended particulates, 0.6 mg/m³ (24-hour averages) [23]. The reported adverse health effects included 6% excess mortality, 12% excess hospital admissions, and 7% excess outpatients. "Excesses" were determined by comparison with a control area in the same region but with low levels of air pollution; evidently the meteorological conditions creating the episode extended over much of Europe but levels of air pollution were governed mainly by local emissions.

Comparison of Episodes

The worst episodes in terms of percentages of people affected, the Meuse and Donora, were recognized as unfortunate incidents when they occurred, but not until impacts on major cities were identified, i.e., London and New York, were air pollution control efforts really taken seriously. The nature of the problem is that, even in a serious episode, the dimensions of the problem, i.e., the number of "excess" deaths, may not be not apparent until detailed statistical analysis has been done later. The recent tragedy in India involving extremely toxic gas was a notable exception.

Even though the worst episodes occurred many years ago and the air pollution levels recorded then are unlikely to ever be experienced again (at least in the developed countries), statisticians are still grappling with the issue of trying to separate the effects of SO_2 from those due to smoke (particulates). In most cases, the worst days featured high levels of both pollutants, making the disentanglement difficult.

Figure 2 plots the excess mortality for the various episodes discussed above (including some additional events for London and New York), versus smoke (Fig. 2a) and versus SO_2 (Fig. 2b). Logarithmic scales are used because of the large

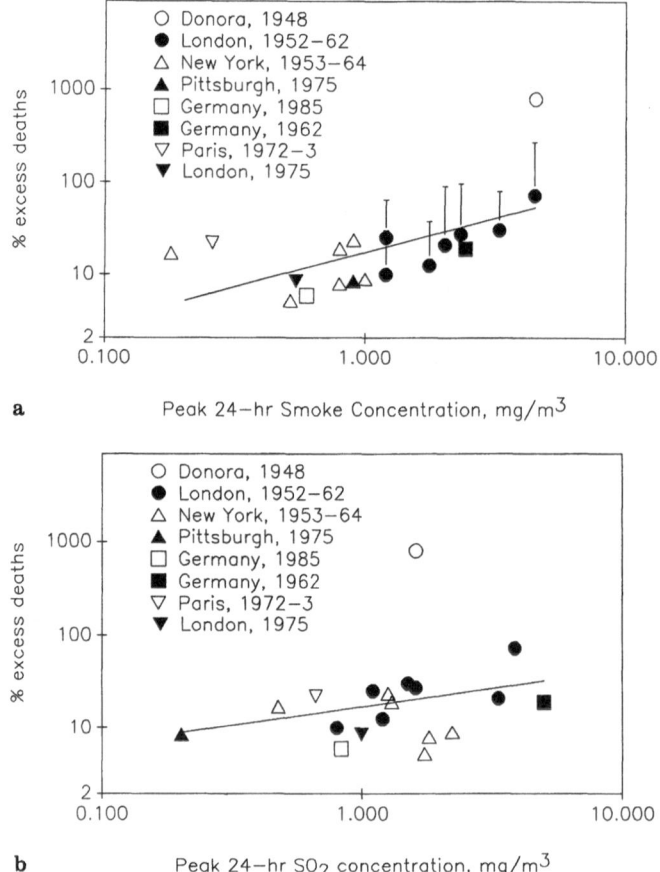

Fig. 2a, b. Episode excess mortality. **a** Percent excess deaths vs. smoke concentration. **b** Percent excess deaths vs. SO_2 concentration. Error bars represent the peak excess mortality day; symbols represent the average excess mortality for the episode

ranges in the variables. The relationship appears to be more consistent in Fig. 2a when plotted against the peak episode smoke concentration. The graph suggests a leveling off or threshold somewhere below about 500 $\mu g/m^3$, but this conclusion is highly dependent on the lowest two data points. It is possible that temperature or other confounding factors played a role in the excess mortality in these two instances. The Donora episode is seen to be an outlier compared to the others; explanations may include the fact that air pollution levels were estimated, metal particles were involved, or that the percentage of excess mortality is uncertain because of the small absolute numbers involved.

Table 1. Regression Models for Air Pollution Episodes

A. Logarithmic models (all available observations)

$$\text{Log (\% Excess mort.)} = 0.63 + \textbf{0.84} \log(\text{smoke}) - 0.2 \log(SO_2) \qquad (1)$$
$R^2 = 0.35, n = 19 \qquad (0.42)^* \ (0.32)^{**} \qquad\qquad (0.37)^{**}$

$$\text{Log (\% Excess mort.)} = 0.50 + \textbf{0.73} \log(\text{smoke}) \qquad (2)$$
$R^2 = 0.33, n = 19 \qquad (0.41) \ (0.25)$

$$\text{Log (\% Excess mort.)} = 0.40 + \textbf{0.81} \log(SO_2) \qquad (3)$$
$R^2 = 0.32, n = 20 \qquad (0.53) \ (0.28)$

B. Logarithmic models without Donora and the Meuse

$$\text{Log (\% Excess mort.)} = 0.76 + 0.37 \log(\text{smoke}) + 0.04 \log(SO_2) \qquad (4)$$
$R^2 = 0.25, n = 18 \qquad (0.27) \ (0.23) \qquad\qquad (0.25)$

$$\text{Log (\% Excess mort.)} = 0.78 + \textbf{0.39} \log(\text{smoke}) \qquad (5)$$
$R^2 = 0.24, n = 18 \qquad (0.26) \ (0.17)$

$$\text{Log (\% Excess mort.)} = 0.86 + 0.29 \log(SO_2) \qquad (6)$$
$R^2 = 0.11, n = 18 \qquad (0.28) \ (0.20)$

C. Linear models (without Donora and the Meuse)

$$\text{\% Excess mortality} = 5.32 + \textbf{12.1} \text{ smoke} - 2.0 \ SO_2 \qquad (7)$$
$R^2 = 0.63, n = 18 \qquad (10.3) \ (2.8) \qquad\quad (2.6)$

$$\text{\% Excess mortality} = 3.99 + \textbf{10.7} \text{ smoke} \qquad (8)$$
$R^2 = 0.61, n = 18 \qquad (9.8) \ (2.1)$

$$\text{\% Excess mortality} = 10.5 + 5.1 \ SO_2 \qquad (9)$$
$R^2 = 0.17, n = 18 \qquad (14.8) \ (2.8)$

D. Linear models without Donora, the Meuse, and London 1952

$$\text{\% Excess mortality} = 11.1 + \textbf{6.4} \text{ smoke} - 2.0 \ SO_2 \qquad (10)$$
$R^2 = 0.33, n = 17 \qquad (6.9) \quad (2.4) \qquad\quad (1.8)$

$$\text{\% Excess mortality} = 9.8 + \textbf{4.9} \text{ smoke} \qquad (11)$$
$R^2 = 0.27, n = 17 \qquad (7.0) \ (2.1)$

Smoke and SO_2 measured in mg/m^3
* Standard error of estimate
** Standard error of regression coefficient
Bold = statistically significant ($p < 0.05$)

Multiple regression analysis was also used to examine the effects of SO_2 and smoke, both separately and jointly. These calculations were done with and without various outliers, for both linear and logarithmic models. The results are given in Table 1. With the log models, the smoke effect appears to be substantially less than linear (slope < 1) when the most severe episodes are excluded. For the linear models, the data for Donora and the Meuse were excluded, because they are highly uncertain and would unduly dominate the linear regressions; the 1952 London episode was excluded in part D of the Table because of the uncertainty in the smoke readings. Smoke appears to be the most important pollutant for these episodes, but the degree of linearity of the relationship depends on which events are included. The linear regression coefficient for smoke is about 11 percent excess deaths per mg/m^3 of smoke when the 1952 London episode is included, but drops to about half that value without this event. The duration of the episode is another potentially important factor to consider.

An additional consideration when comparing episodes is the degree of uniformity of exposure of the population at risk. One could speculate that exposure may have been more uniform in the smaller cities (Donora and the Meuse) and perhaps in London in 1952 (because of the widespread use of soft coal for home heating), than in New York City in the 1960s, for example. If so, the higher responses seen in these three cases would more nearly represent the "true" effects. One must also consider that there may be effects from unmeasured air pollutants, such as carbon monoxide, which could influence the slopes and intercepts of the regressions.

Daily Mortality Time-Series Studies for Longer Periods

Figure 2 provides evidence of the *existence* of effects of air pollution on human mortality; the similarity of mortality responses from several different times and places seems to rule out substantial confounding by meteorological variables or by faulty reporting (but not by covarying pollutants, which may be similar in various cities). However, these data do not provide very useful information for the establishment or confirmation of ambient air quality standards, since most of the data relate to pollution levels well above those suitable for standards intended to protect public health. For this purpose, analyses of time-series of daily mortality (or morbidity) and pollution are required. More studies have been done on mortality than morbidity since mortality data are more readily available.

Since weather affects both air pollution and health [26, 27], it is important to account for these interactions correctly at the lower pollution levels where effects may be more subtle. During approximately the past decade, much of the literature on short-term health effects has been concerned with this topic.

Figure 3 is a flow chart which may be useful, in the context of path analysis. Beginning at the top left of the chart, weather patterns are seen to influence both the emissions of air pollution and their dispersion (paths W–E and W–C), both of which are responsible for daily fluctuations in concentrations of air pollutants.

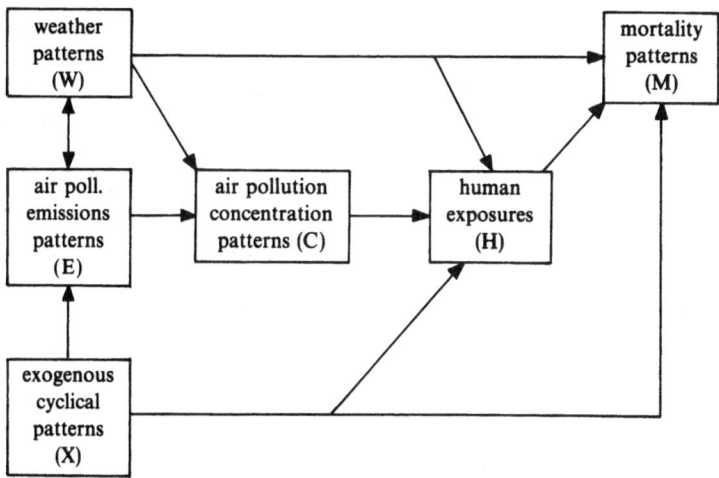

Fig. 3. Causal path diagram for mortality time-series analysis

Research in biometeorology [28] suggests that weather patterns can have a direct influence on health and mortality (path W–M) and can also influence the patterns of human that control exposure to outdoor air pollution (path W–H). Thus, to adequately estimate the importance of path C–H–M, which is the goal of all of these analyses, one must consider both of the other paths (W–M and X–M) influencing mortality patterns. The latter includes not only the well-known seasonal patterns, which affect mortality regardless of the weather or climate (path X–M), but also include effects of holidays and the day of the week. The various studies of mortality-air pollution time-series vary greatly in the ways they account for these potential confounding cycles.

Studies of Daily Mortality in London

Early Studies

The effects of air pollution and fogs on health in London have been studied for over 60 years. Russell [29] analyzed a 21-year record beginning in the 19th century, using regression and correlation analyses of weekly deaths on fog and temperature variables. His technique involved numbering the weeks of each year and examining the relationship of respiratory mortality for each week across the 21 years, for two residential boroughs of London. He concluded that fog and cold weather had a larger effect than fog alone, and that usually only adults were affected. Because emissions from space heating are increased during colder temperatures, an air pollution effect could also be implied.

Logan [30] analyzed the causes of death attributed to polluted fogs, with reference to episodes occurring during the winter of 1948. He also found that the excess deaths occurred in people aged 45 and older, primarily for bronchitis and pneumonia with some contributions from cancer and myocardial degeneration (but not "old age").

Early mortality analyses in London that incorporated air pollution data include the work of Gore and Shaddick [31], Martin and Bradley [32], and Martin [33]. All of the major studies of London mortality have used a data base comprising the averages of seven stations for British smoke and SO_2, beginning with Gore and Shaddick [31]. These data are reported for the 24-hour period ending at 9 A.M., and thus represent nearly a 1-day lag with respect to the mortality counts reported for the same day. The report of Martin [33], dealing with the winter of 1958–9, does not specify the source of air quality data, described as "indices" of smoke and SO_2 in Greater London.

Gore and Shaddick [31] provide some descriptions and a map of the 7-station network (Fig. 4), which was established at the end of 1954 to measure smoke and SO_2 on a continuous basis, using a sequential sampler which allowed six days' unattended operation. The sites consist of the County Hall (presumably on the roof), centrally located in Westminster, and six ambulance stations, distributed throughout London County. Gore and Shaddick state that all samples were taken from the same height and that none of the monitors had been established "to measure a particular local source of pollution". They also noted that the prevailing winds are from the southwest. Some information is also available from various published sources on levels in London of aerosol acidity, sulfates, carbon monoxide, and trace metals.

Figure 5 compares SO_2 and smoke readings from this network and several other measuring sites [34, 35], located in the 2.6 km^2 part of London called "The City". The figure shows how the network smoke readings in winter diverge from the "City" levels before about 1962, apparently because of the lack of smoke control in outlying areas. Within central London, the differences between "St. Barts" and

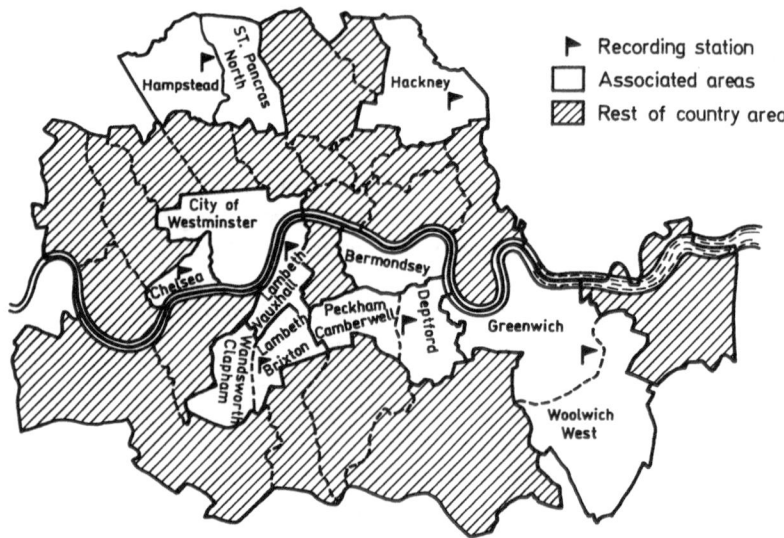

Fig. 4. Map of the 7-station air pollution monitoring network in London County. Source: Gore and Shaddick (1958) [31]

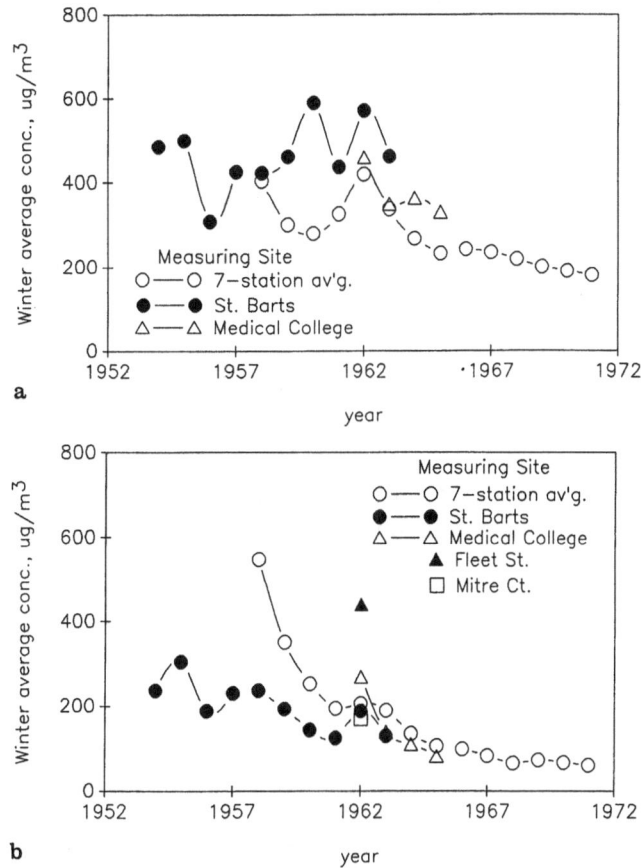

Fig. 5a, b. Comparison of London air pollution trends. **a** SO$_2$. **b** Smoke. Data from Refs. [34] and [35]

"Medical College" reflect gradients over a distance of about 600 m [34]. The "Fleet St." point was taken during a study of traffic pollution [35] with sampling at 1.2 m height from a traffic island in the street. The "Mitre Ct." point was sampled from the fourth floor of a building 46 m away from the traffic sampler. All of the data from the traffic study were averaged from 8 A.M. to 7 P.M., Mondays to Fridays, for a year (divided into summer and winter portions). Carbon monoxide at the Fleet St. site averaged 7.8 ppm continuously, with an average of 16.6 ppm for weekdays, 8 A.M. to 7 P.M. Hourly plots at this site show a nearly continuous plateau of high concentrations from about 9 A.M. to about 7 P.M. At the St. Barts laboratory, CO never exceeded about 2 ppm [36].

Gore and Shaddick studied four episodes during 1954–6 but did not perform a time-series analysis *per se*. These episodes were selected on the basis of high pollution readings at all seven of the network stations in conjunction with foggy conditions; apparently this combination of conditions is then referred to as a "fog". Their data show the influence of both smoke and SO$_2$ on excess mortality for

London County, and they identified 24-hour average levels of 2.0 mg/m^3 smoke and 1.1 mg/m^3 SO$_2$ as the approximate onset of marked increases in mortality. They used seven-day moving averages for this analysis, so that "harvesting effects" (mortality decreases immediately after an episode) would have been smoothed out. However, as pointed out by Holland et al. [19], mortality increases were seen in their data at levels as low as 1000 and 750 μg/m^3, respectively. Gore and Shaddick noted that the worst of these four events hit older people with bronchitis particularly hard, whereas cold periods (without "fog") produced increases in cardio-vascular deaths. They also studied the relationship between long-term mortality in certain areas and pollution levels as recorded by the individual stations of the network (discussed in more detail below).

Martin and Bradley [32] analyzed the mortality data from Greater London for the winter of 1958–9, using mortality deviations from a 15-day moving average as the dependent variable. This technique removes the effects of long-term variations, for example, due to weather or season. Pollution data were based on the seven-station network for London County, although reference was also made to a 12-station network for Greater London which usually had lower readings. (This study thus apparently established the precedent of using the London County pollution data in conjunction with Greater London mortality data, even though the latter area was about five times bigger and had 2.6 times the population.) For 1958–9, Martin and Bradley found statistically significant correlations between either smoke or SO$_2$ and mortality from all causes and from bronchitis, but apparently without jointly considering the effects of daily temperature changes (temperature and relative humidity were only weakly correlated with mortality). The relationship was slightly stronger for smoke and for a model using the logarithm of the pollutants, but was not noticeably improved by considering smoke and SO$_2$ jointly. A log model provided no better fit for the episodes shown in Figs. 2 and 3, when percent excess deaths (rather than the log of percent excess deaths) was chosen as the dependent variable.

Martin [33] extended the technique to the winters of 1958–69. Regression analysis of the tabulated excess mortality data suggested a threshold for smoke at 450 μg/m^3 and 330 μg/m^3 for SO$_2$ (considered separately without temperature effects), although Martin commented on the difficulty of defining "safe" levels of pollution. The slopes of these regressions were 13% excess mortality per mg/m^3 smoke, or 16% excess mortality per mg/m^3 SO$_2$, which are somewhat higher than those derived from the multicity episodes (Table 1).

Macfarlane [36] discussed the importance of proper consideration of influenza epidemics and concluded on the basis of visual inspection that there was no longer an association between daily mortality and suspended particulates, based on the 7-station network (ca. 1970). He also commented that in 1965 the study area was changed slightly and data collection was begun on a year-round basis, and that the codings for causes of death were changed in 1968 in accordance with changes in the International Classification of Diseases.

Studies by the Schimmel, Mazumdar, and Higgins Team

Mazumdar et al.'s [37] analysis was the first comprehensive study of a long time-series of Greater London mortality (14 winters, from 1958–72). The environmental data were daily mean values of SO_2 and British (black) smoke, averaged over the seven London County measuring stations, and readings of temperature and relative humidity taken at 9 A.M. at a single site. The 15-day moving average was used, dummy variables were included for days of the week, and variables were divided by their mean (winter) values to account for the long-term trends (which were considerable in part because of the air pollution controls that had been instituted; year-to-year changes such as discussed by Macfarlane [36] would also be accounted for by this procedure). Weather effects were handled through a two-stage procedure: first, mortality and pollution variables were regressed on temperature and humidity to provide "corrections". The corrected variables were then analyzed for the pollution-mortality relationships. The authors reported that the results of this two-stage procedure did not differ substantially from those obtained using weather variables jointly in the same regression with pollution and mortality.

Mazumdar et al. [37] made several types of analyses. First, they provided regressions for each winter, and averages for the first and second groups of seven and for all fourteen as a group. Smoke and SO_2 were regressed against mortality separately and jointly; a separate set of mortality regressions was also provided for $(smoke)^2$, $(SO_2)^2$, and $(smoke \times SO_2)$ to assess the linearity of responses and interactions. Additional yearly regression results were obtained from their progress report [9]. Twelve of the fourteen individual SO_2 regressions and thirteen of the individual smoke regressions were significant. Six of the fourteen joint regressions had significant smoke coefficients (positive): of the remaining winters, two had significant (positive) SO_2 coefficients. The 1964–5 winter had a significant positive smoke coefficient and a significant negative SO_2 coefficient. For the first seven winters (mean smoke from 0.13 to 0.55 mg/m^3; mean SO_2 from 0.27 to 0.42 mg/m^3), the overall SO_2 coefficient was -3.3 and the smoke coefficient, 22 (% excess deaths per mg/m^3). For the second group (mean smoke from 0.06 to 0.11 mg/m^3; mean SO_2 from 0.18 to 0.24 mg/m^3), the overall smoke coefficient was 28 and the SO_2 coefficient, 5.6. For all fourteen winters, the smoke coefficient was 25 and the SO_2 coefficient, 1.2 (all in units of % excess deaths per mg/m^3). None of these three grouped SO_2 coefficients was statistically significant; all of the smoke coefficients were significant. Note that the mortality regression coefficients for smoke for these comprehensive monitoring periods tend to be substantially larger than those estimated from episodes (Table 1).

Mazumdar et al. [37] introduced the parameter, "mortality effect", obtained by multiplying the regression coefficient by the corresponding mean value for pollution, expressed as a percentage of mean mortality. Figure 6a compares the effect on mortality as expressed, alternatively, by both pollutants and by each separately, plotted against the annual mean smoke concentration. Best-fit second-order polynomial regression lines are also shown (since the trend is not linear, contrary to expectations) and although there are some differences for individual years, the best-fit lines are virtually identical. The inability to distinguish between the two pollutants stems from their high correlations for the daily values, which ranged from 0.79 to 0.96. The analysis indicates a pollution effect on mortality of about 3% at the lowest pollution levels, rising to almost 9% for the worst year (1958–9). In Fig. 6b, the mortality effect data are plotted against their respective mean pollutant concentration levels: SO_2 effect versus SO_2 concentration, smoke effect versus

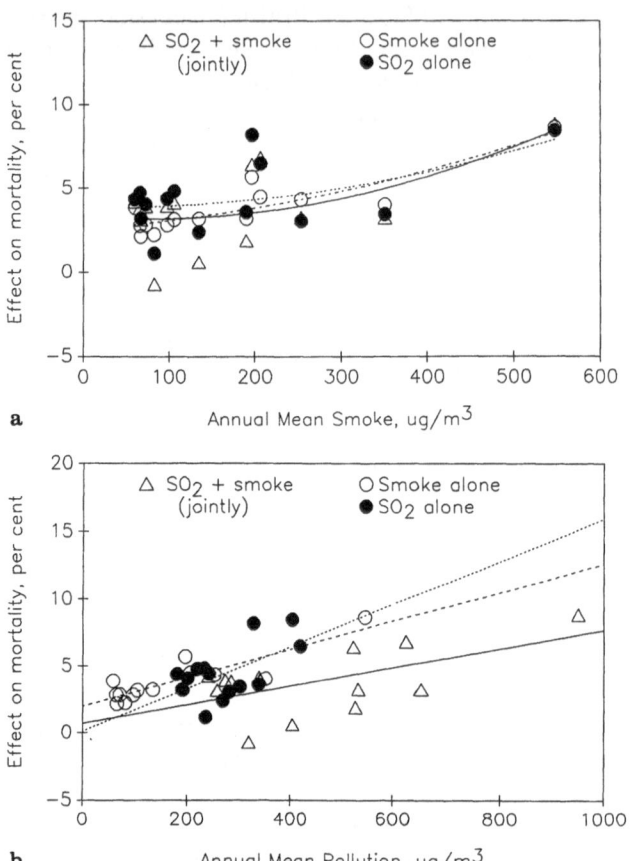

Fig. 6a, b. Comparison of mean pollution effects on mortality. **a** Plotted vs. smoke. **b** Plotted vs. various combinations of smoke and SO_2. Data from Mazumdar et al. (1982) [37]

smoke concentration, and the joint effect versus the sum of SO_2 plus smoke concentrations. The figure suggests that the effect of pollution on mortality only passes through the origin when SO_2 is considered as a contributing pollutant.

Mazumdar et al. addressed the question of separation of the effects of the two pollutants in three different ways: the joint regression, for which SO_2 was usually not significant; a nested quartile analysis; and a case study of high pollution days. Based on the nested quartile analysis, the authors report a coefficient of 22% per mg/m^3 for smoke and 6% for SO_2. The analysis of high pollution days yielded a model with both positive and negative terms for the linear and squared pollution terms, indicating that SO_2 was important at the higher concentration levels. On the basis of these three analyses, Mazumdar et al. concluded that smoke was much more important than SO_2 in London mortality and that it likely had a non linear dose-response relationship.

This view was challenged by Goldstein et al. [38] on grounds of the methodology used by Mazumdar et al., which they felt was overly complicated and unjustified. They also criticized the neglect of possible confounding by influenza epidemics, the emphasis on crude total mortality, and the arbitrary deletion of outliers. In their response to this criticism, Mazumdar et al. amplified their findings with additional tables of year-by-year regression results [39]. First, they supplied R^2 values for the regressions based on two-stage temperature corrections. For the joint regressions on SO_2 and smoke, these values were all in the range 0.16 to 0.38 except for the two winters (0.57 for 1958–9 and 0.70 for 1962–3) which had statistically significant coefficients for SO_2 (but not for smoke). The year with a significant negative SO_2 coefficient had an R^2 of 0.29. Next, they presented results for regressions of mortality on SO_2 and smoke, as deviations from 15-day moving averages, without considering weather or day of the week. The SO_2 effects were decreased slightly and the smoke effects increased; R^2's decreased for all years. The smoke coefficient increased for the later years, which had low mean values of smoke. The estimated average mean effect of smoke on mortality increased slightly over the "corrected" values. The final table of regression results was based on "crude" variables with no seasonal adjustments; these results were quite similar to their previous results, except that R^2's were uniformly low and the smoke effects were increased slightly. The year-to-year variation in smoke coefficients was greatly increased and had little resemblance to the other results for specific years; however, the average for the 14 years was not greatly different. In their discussion of "outliers", Mazumdar et al. presented mortality and pollution data for the five days of the December 1962 episode, and point out how their model greatly overpredicts the maximum mortality actually observed during this event. However, the slope of mortality on pollution for these five points is significant for SO_2 with a value close to that obtained for the entire winter, while smoke is nonsignificant. This finding suggests that SO_2 played a role in London mortality, in spite of the consistent tendency for multiple regressions to indicate only smoke.

Mazumdar et al. tabulated the nested quartile data using temperature corrections [35] and on several different bases [9]. Since each of the 16 quartiles contains the same number of original observations, these data provide a convenient means of reanalyzing the London data set for the purposes of this chapter. The quartiles were intended to separate the effects of SO_2 and smoke, by alternatively holding each constant (nesting). This procedure was only partially successful, since the correlation between the two was still 0.89 for the 16 quartiles. However, if only the quartiles are used for which one pollutant is held reasonably constant while the other is varied, the resulting smoke coefficients range from 12% to 36% and the SO_2 coefficients from -9% to 19% per mg/m^3.

A joint regression of the tabulated quartile data from Ref. [9], with corrections for temperature, yielded the model (standard errors of regression coefficients in parentheses):

$$\% \text{ excess mortality} = 0.0064 + 8.0\, SO_2 + 11.9 \text{ smoke} \qquad (12)$$
$$R^2 = 0.85 \qquad\qquad (4.3) \qquad (4.8)$$

Dropping the highest smoke quartiles resulted in:

$$\% \text{ excess mortality} = -0.013 + 12.0\, SO_2 + 6.8 \text{ smoke} \qquad (13)$$
$$R^2 = 0.74 \qquad\qquad (4.0) \qquad (6.8)$$

The residuals from these two models are plotted against smoke in Fig. 7; the differences between the two are quite small, which illustrates the difficulty in partitioning the effects of the two pollutants. The data at about 250 $\mu g/m^3$ smoke with a residual of about 2% have a large influence on this analysis. Figure 8 shows the dose-response relationships of each pollutant alone. These plots suggest a threshold at about 100 $\mu g/m^3$ (0.1 mg/m^3), but these aggregated data do not permit statistical testing of thresholds.

Reference [9] presents tabulations of nested smoke and SO_2 quartiles based on deviations from 15-day moving averages, for both "episodic" and "nonepisodic" periods. As defined by Mazumdar et al., an episodic period is a day with smoke exceeding 0.5 mg/m^3 and the seven days preceding and following. To combine

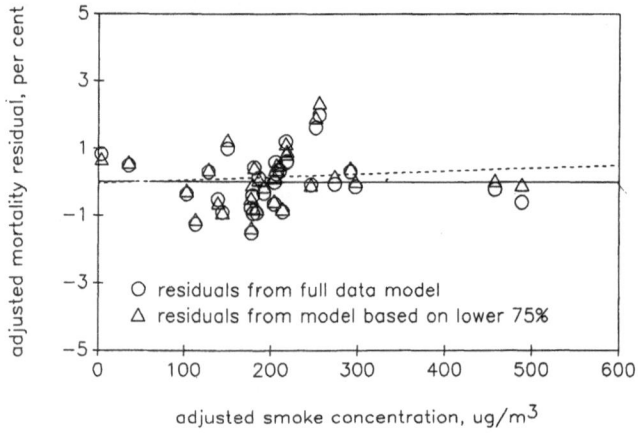

Fig. 7. Residuals from models for London mortality quartiles, winters, 1958–72. Data from Mazumdar et al. (1982) [37]

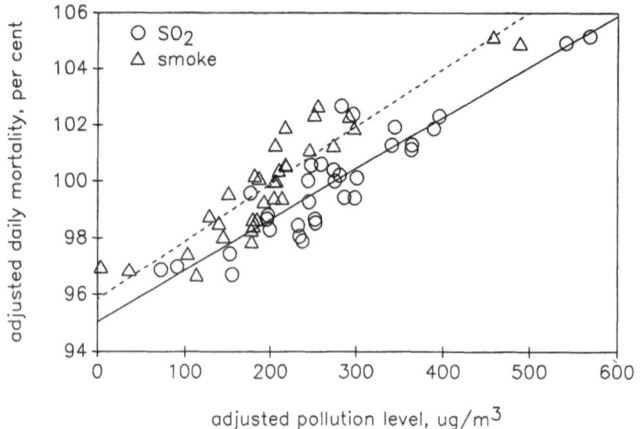

Fig. 8. Dose-response relationships for London mortality quartiles, winters, 1958–72. Data from Mazumdar et al. (1982) [37]

these two data sets, mortality data must be converted back from percentages to daily death counts, since the two periods were normalized by different mean values. Figures 9 to 11 present these data. Figure 9 plots the dose-response relationship for smoke as the sole pollutant. The nonepisodic data have a significantly higher slope (88 versus 57 deaths/mg/m^3), giving rise to the logarithmic or quadratic relationship that others have noted. Figure 10 shows the same data plotted against SO_2. The slopes of the two data sets are almost the same, at 52 and 56 deaths/mg/m^3. The difference between the two lines is due to the difference in average smoke level (249 $\mu g/m^3$), plus all other differences between episodic and nonepisodic periods (such as fog and cold). A detailed view of the nonepisodic period is given in Fig. 11. Again, there is the suggestion of a threshold at about 100 $\mu g/m^3$. Using linear

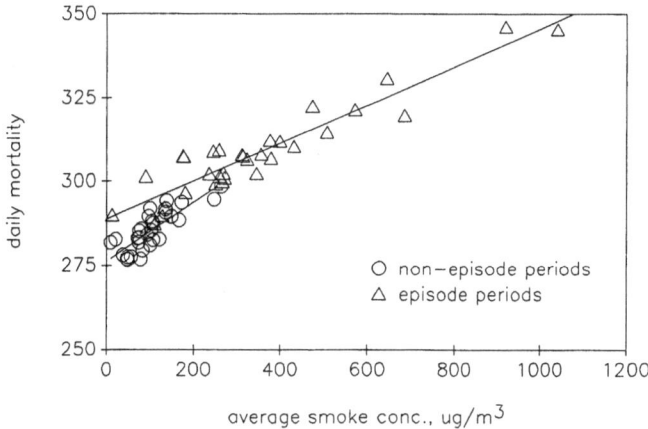

Fig. 9. Dose-response relationship for London mortality quartiles, plotted vs. average smoke concentrations, winters, 1958–72. Data from Mazumdar et al. (1982) [37]

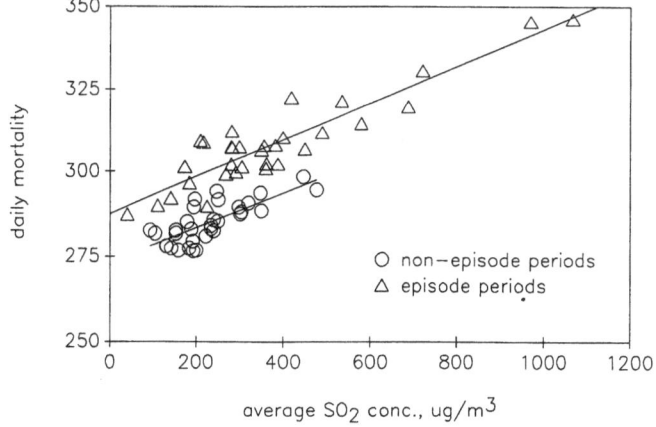

Fig. 10. Dose-response relationship for London mortality quartiles, plotted vs. average SO_2 concentrations, winters, 1958–72. Data from Mazumdar et al. (1982) [37]

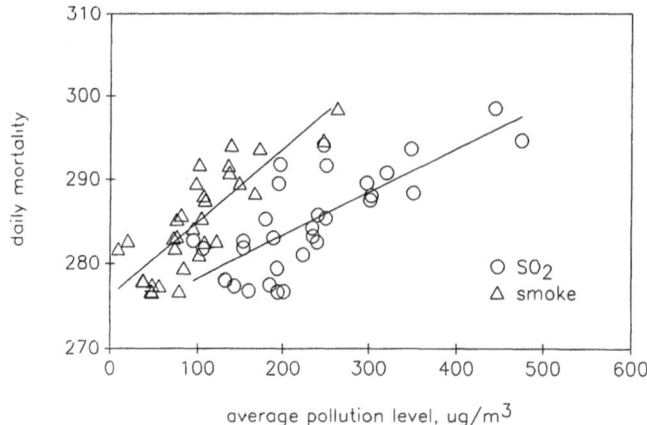

Fig. 11. Dose-response relationships for non-episodic London mortality quartiles, winters, 1958–72. Data from Mazumdar et al. (1982) [37]

models, the joint regressions for these data are (standard errors of regression coefficients in parentheses):

For episodic periods:

$$\% \text{ excess mortality} = -0.066 + 3.9 \, SO_2 + 14.6 \, \text{smoke} \qquad (14)$$
$$R^2 = 0.88 \qquad\qquad\qquad (3.6) \qquad\quad (4.5)$$

For nonepisodic periods:

$$\% \text{ excess mortality} = 0.03 + 0.0 \, SO_2 + 31.3 \, \text{smoke} \qquad (15)$$
$$R^2 = 0.68 \qquad\qquad\qquad (3.6) \qquad\quad (10.2)$$

For both periods combined:

$$\% \text{ excess mort.} = -0.05 + 5.5 \, SO_2 + 14.6 \, \text{smoke} + 3.6 \, E \qquad (16)$$
$$R^2 = 0.93 \qquad\qquad\quad (3.6) \qquad\quad (3.6) \qquad\qquad (0.57)$$

where E is a dummy variable for episodic versus nonepisodic periods. Smoke is always statistically significant, while SO_2 only approaches significance in Eq. (16) ($p > 0.10$).

To explore these relationships further, the residual patterns were checked for evidence of heteroscedasticity, which was not apparent. In addition, statistical tests were performed to see if the degrees of fit provided by using one or both pollutants [Eq. (12)] were significantly different. This analysis showed that adding SO_2 made a significant improvement in fit to a model with smoke alone, even though the coefficient for SO_2 was not quite statistically significant.

The final attempt at rederiving dose-response relationships from Mazumdar et al.'s studies of the London mortality data involves their tabulated means of the data by day of the week (Fig. 1), for episodic and nonepisodic periods. Again, the best regressions were all based on smoke; adding SO_2 made little difference. Smoke coefficients for the combined period were in the range 27–32 percent excess deaths per mg/m^3.

Subsequent Studies of the 1958–72 London Mortality Data Set

These data were used to address the question of ambient air quality standards in the United States. Ostro [40] used deviations from 15-day moving averages with temperature and humidity entered directly into the regression model and found "... no evidence to support the existence of a no-effects level. Further, the reanalysis suggests that the estimated pollution-mortality relationship exists even in nonepisodic winters, when British Smoke (BS) readings were less than 500 $\mu g/m^3$". For days with smoke greater than 150 $\mu g/m^3$, Ostro found an average coefficient of 49 deaths/mg/m^3; for values less than 150, an average of 146 deaths/mg/m^3. This finding is qualitatively consistent with the nonlinear dose-response relationship found by Mazumdar et al., discussed above (Fig. 9). However, such a dose-response function has a counterintuitive shape (the lower the pollution concentration, the more lethal it is!). An alternative explanation is that important factors were omitted from the analysis and that at low levels, smoke is behaving as a surrogate for some other factor, possibly SO_2 or CO, among others. Ball and Hume [17] showed a strong correlation (r = 0.98) between black smoke and particulate lead at a non-traffic site in London, raising the possibility that other motor vehicle air pollutants may be similarly correlated.

Additional criticisms of the Ostro study include the omission of variables for flu epidemics, holidays, and days of the week. In addition, more rigorous statistical methods for searching for thresholds employing dummy variables could have been used.

The next analysis of this London mortality data set was performed by Schwartz and Marcus [41], who used both autoregressive models (allowing for correlations between data on successive days) and deviations from a 15-day moving average. They also performed regressions with and without temperature and humidity variables in the regression model and explored subsets defined by smoke concentration level. The authors found that "... multiple regression models for daily mortality and BS do indeed reflect a relationship that cannot be attributed to time-series effects, temperature, SO_2, or functional misspecification .... There does appear to be some tendency for higher slopes in later years when both BS and SO_2 reflect more nearly contemporary conditions." This study established the autocorrelation structure of the data and showed that deaths on day t are correlated with deaths on day t − 1, even when deviations from a 15-day moving average are used. The dose-response findings generally confirmed previous ones: smoke coefficients increasing in the later years as mean levels dropped and smoke positive with SO_2 negative in joint regressions. However, Schwartz and Marcus also showed that including temperature and humidity resulted in larger and more significant pollution coefficients, especially for SO_2. When SO_2 was used as the sole pollutant in a random-effects model across all years, the result was substantially more significant than when smoke was used alone (t = 21 versus t = 6). Schwartz and Marcus concluded that an independent SO_2 effect on mortality could not be excluded, but that the effects of smoke were independent of SO_2.

Shumway et al. [42] emphasized spectral analysis in their analysis of the London data. They also separately examined the daily counts of total, cardiovascular, and respiratory mortality. The authors concluded that the best model uses temperature in conjunction with the logarithm of either smoke or SO_2. The two pollutants appear to be acting identically in all respects; no threshold effect was evident. Relative humidity was not found to be important. The strongest coherence occurs at periods of 7–21 days, implying that pollution or temperature episodes must be longer than 7–21 days to have a discernible effect on mortality. Confirming earlier findings, they showed that temperature effects were more important than pollution for cardiovascular mortality, but not for respiratory mortality. They warned that these results should not be extrapolated to the United States, a point that was reinforced by Roth et al. [43] on the basis of differences in indoor air quality and the chemical composition of particulate matter.

With respect to the threshold question, Shumway et al.'s analysis is silent on year-to-year variations during the period of improving air quality since the data for each year were normalized with respect to the mean values for that year. Shumway et al. used logarithmic transforms for the pollution variables and thus effects of the temporal changes in mean pollution values which occurred over the years would tend to be suppressed. Thus the authors' conclusion: "No threshold effect was evident." should not be interpreted as providing evidence that thresholds do not exist since the study was not structured to find them.

The report's conclusion regarding the ineffectiveness of short-period (< 7 days) episodes is also curious. The best evidence for short-term mortality effects of air pollution comes from severe episodes typically lasting 3–4 days [30–33], during which the excess mortality was quite obvious. Since there were probably only a few of those episodes during the period studied, the results may be weighted more towards periodic excursions in weather and the accompanying effects on pollution. A consistent dose-response relationship over the years when mean pollution levels were changing greatly would be the best evidence that the excess mortality is in fact caused by air pollution (i.e., smoke or SO_2) and not by temperature. If this were in fact the case, air pollution would be a coincidental rather than a causal variable. Other (unmeasured) pollutants such as carbon monoxide cannot be ruled out. Therefore, it may be useful to analyze the London data on a continuous year-round basis, rather than just during the winters.

Comparison and Reconciliation of the London Studies

The most important issues remaining from the various analyses of the London daily mortality data are concerned with the shape of the dose-response relationship(s) and the separation of smoke effects from those of SO_2. In the analysis which follows, these two issues and their interrelationship will be examined.

The changing pollution levels over the fourteen winters provide an opportunity to examine dose-response relationships on an aggregated basis. First, since the daily mortality rate decreased by 25% during period of study, for reasons which are only partly related to air pollution, it is important to calculate regression coefficients as *percent* excess deaths per unit of pollution, as opposed to death counts (an alternative would be to account for the secular change separately, for example, by pooling the data and including an appropriate independent variable in the regression). If pollution is measured in milligrams per cubic meter, all the coefficients will be in the units used by Mazumdar et al. [37]. It is also useful to consider some theoretical dose-response models, as shown schematically in Fig. 12, in terms of both a regression coefficient and the corresponding effect on mortality.

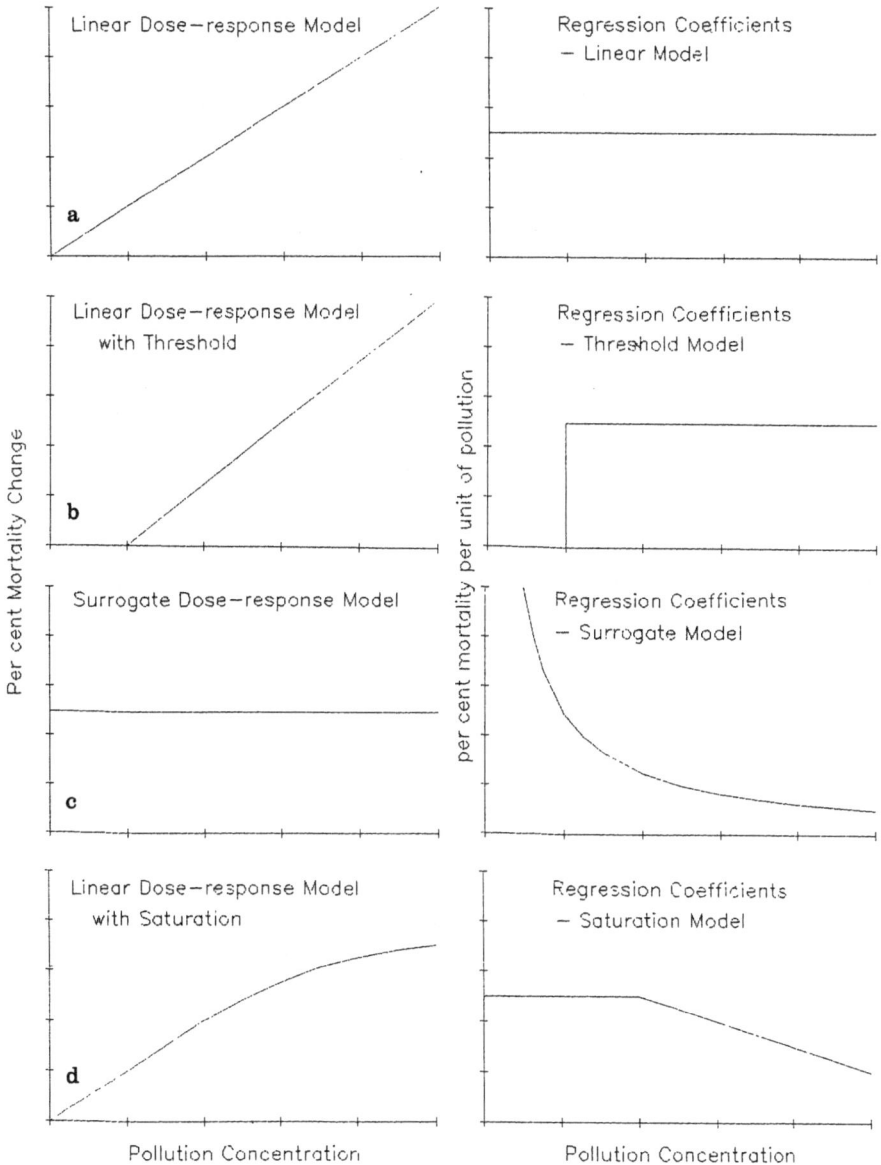

Fig. 12a–d. Theoretical dose-response models. **a** Linear, no-threshold model. **b** Linear threshold model. **c** Surrogate model. **d** Linear model with saturation. The corresponding regression coefficients (slopes) are shown to the right of each model

A linear, no-threshold, model (Fig. 12a) has a constant coefficient (C) over the entire range of pollution (P) and an effect (C × P) on mortality which is linear and passes through the origin. A linear threshold model (Fig. 12b) is similar except that the coefficient becomes indeterminate between zero and the threshold pollution level (P_0, and the mortality effect has an x intercept at P_0. The next possibility (Fig. 12c) is the surrogate model, which is characterized by a constant effect (or an effect not influenced by the pollutant used for plotting and analysis), and therefore has a coefficient which increases with decreasing pollutant concentration as the (constant) effect is divided by smaller values. Nonlinear models (Fig. 12d) in theory may be either have coefficients which increase with pollution level, because effects become more severe at higher exposure levels, or decrease, because saturation is reached and the dose loses effectiveness. The latter case is shown in the figure.

Figure 13 is a plot of SO_2 coefficients (as the sole pollutant) as determined from the London data by Schwartz and Marcus (S&M) [41], with and without consideration of temperature and humidity effects ("t & h"), and by Mazumdar et al. [37]. Below about 300 $\mu g/m^3$, the coefficients generally are no longer significant, although the estimated coefficients remain remarkably constant over the range of annual mean SO_2 levels. Linear regression lines are shown to depict the relationships between coefficients and mean SO_2 levels; none have slopes significantly different from zero, suggesting that the regression coefficients are independent of the mean pollutant level (Fig. 12a). Since all three of these analyses produced virtually indistinguishable results, it appears that temperature/humidity corrections are not critical. The corresponding plot of mortality effects, Fig. 14, passes through the origin, as in Fig. 12a. However, if the lower confidence limits on Fig. 13 had been used to compute the mortality effect, a threshold of effect would have been shown as in Fig. 12b. Such a threshold is suggested from the joint regression for SO_2 and smoke; the SO_2 contribution approaches zero at $SO_2 = 260$ $\mu g/m^3$.

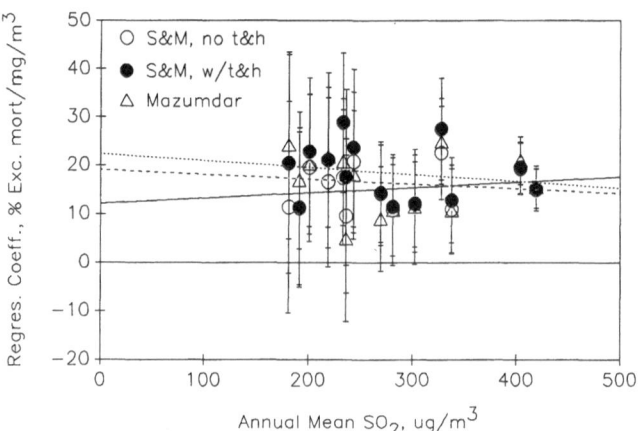

Fig. 13. Comparison of SO_2 regression coefficients for London mortality, winters of 1958–72. Lines represent linear regressions of the individual regression coefficients vs. SO_2 level; error bars are for 2-sigma confidence limits on the individual regression coefficients. Data from Refs. [37] and [41]

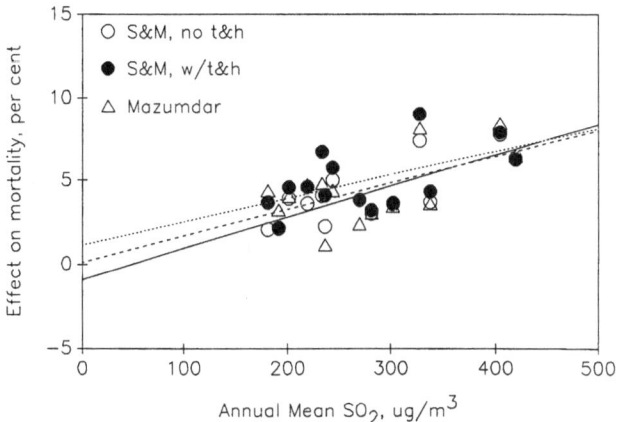

Fig. 14. Comparison of mean SO_2 effects (regression coefficient × mean SO_2 level) on London mortality, winters of 1958–72. Lines represent linear regressions of the individual effects vs. SO_2 level. Data from Refs. [37] and [41]

Figure 15 compares the regression coefficients for smoke. Six different regression models are compared in Fig. 15a; the results are quite consistent above about 200 $\mu g/m^3$ and correspond closely with the results obtained from episodes in several cities (Fig. 2a). Below this level, the curve looks like a surrogate model (Fig. 12c). Confidence limits are shown on Fig. 15b for two of the models presented by Schwartz and Marcus [41]; most of the lower confidence limits would allow a constant coefficient consistent with the episode data (11% per mg/m^3). Figure 16 shows the corresponding smoke effects on mortality, and additional data for Mazumdar et al.'s regression for smoke and SO_2 together. Linear regressions (not shown) through these data indicate a statistically significant mortality intercept of about 2.5–3%, which is a symptom of a surrogate model (Fig. 12c).

Conclusions from Studies of London Mortality

From a statistical point of view, the various studies are quite consistent. Smoke is the most important pollutant, and its effect appears to become more severe per unit as concentrations decreased when clean air controls were imposed. The interpretations of this finding differ considerably. Ostro [40] argues that the lack of a threshold has been demonstrated; however, there is no support from other studies or from clinical exposures that exposure to 24-hour average values of smoke of less than 150 $\mu g/m^3$ could have such severe effects on health. However, Ostro's analysis ignores the simultaneous presence of over 180 $\mu g/m^3$ SO_2 and levels of CO probably up to 15–20 ppm. The dilemma in the analysis is the consistent lack of statistical significance of SO_2 when regressed jointly with smoke. It is clear that this situation results from the high correlation between the two pollutants in the London data base. However, it is also clear that the spatial distributions of the two pollutants are quite different [18, 31], and thus the high correlation may be due partly to averaging across the seven stations of the network. When two independent variables are highly correlated, the variable with the least measurement error

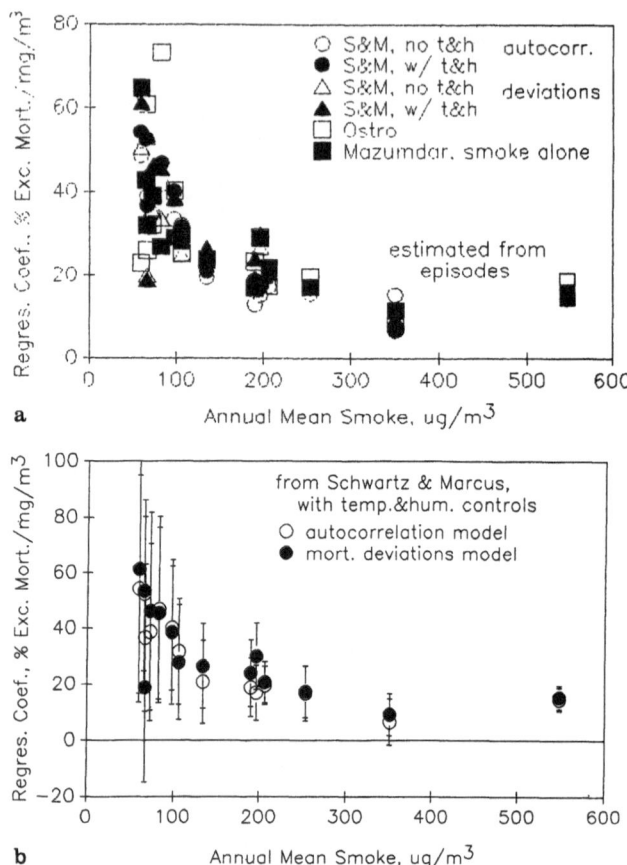

Fig. 15a, b. Comparison of smoke regression coefficients for London mortality, winters of 1958–72. **a** Various models. **b** Models due to Schwartz and Marcus. Data from Refs. [37, 40, and 41]. Error bars in **b** represent 2-sigma confidence limits

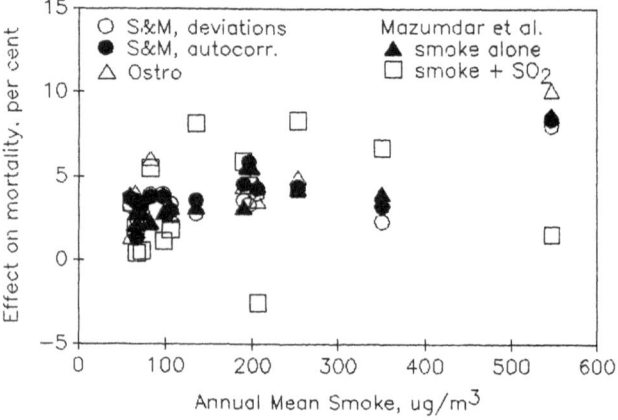

Fig. 16. Comparison of mean smoke effects on London mortality, winters of 1958–72. Data from Refs. [37, 40, and 41]

will dominate the regression. Since the sources of SO_2 in London shifted over the years from distributed area sources to discrete point sources [44], it is possible that the seven monitoring stations represent smoke exposures better than SO_2 exposures. This would explain the existence of a physiological effect of SO_2 in London, despite the lack of statistical significance when regressed jointly with smoke. Effects due to carbon monoxide are another possibility, but lacking data, this hypothesis cannot be explored. My recommendation is that a model such as Eq. (12) or Eq. (16) be used to estimate the effects of air pollution on mortality in London.

Times Series Studies of Mortality in New York City

As in London, studies of the effects of air pollution on mortality in New York began with descriptions of isolated incidents [20, 45] and progressed through a series of more formal time-series analyses. A useful summary was given at a symposium in March, 1978 [46].

Routine air pollution monitoring in New York City began in the 1950s at the 121st Street Laboratory station in upper Manhattan, where SO_2 was measured with the peroxide method and smokeshade by the tape sampler, reporting the readings as "coefficient of haze" (COH). One COH unit could correspond to a range of mass units, from 69 to 200 $\mu g/m^3$ [19]. At high levels, a conversion factor of 100 $\mu g/m^3$ per COH unit is reasonable and convenient [4]. In 1962, Cornell University Medical College established an air monitoring station about 10 km further south on the lower east side of Manhattan, data from which were used in several studies [20, 47]. By the 1970s, an extensive network was in operation in all five boroughs of the city (ca. 40 stations), which has been scaled back considerably in recent years. Goldstein [48] describes some of the details and interrelationships among the stations of this network.

Peak air pollution levels were recorded in New York during the 1960s, a period when residual fuel oil with high (2–3%) sulfur content was used extensively in large commercial and residential buildings. Also, it was the practice to require on-site incineration of municipal refuse, which added to the particulate burden [49]. Hourly values of SO_2 as high as 1 ppm and smokeshade as high as 8.0 COHs were recorded then during adverse meteorological conditions [21]. Carbon monoxide readings in the mid-1970s ranged from 3–15 ppm annual average, with peak 8-hour averages as high as 30 ppm [49]. Readings were probably higher in the 1960s before automotive pollution controls came into use.

Early Studies in New York

The early studies of episodes focused on mortality by cause and age [50, 51] rather than on developing dose-response relationships. Causes which were affected include influenza-pneumonia, vascular lesions of the central nervous system, diseases of the heart, and lung cancer. All age groups were affected, except perhaps infants. The air pollution levels cited above, which occurred in November 1966, were associated with about 10% total excess mortality, for the entire city [50].

Greenburg et al.'s study [51] focused on influenza, which was a major factor in January–February 1963, a time when SO_2 was also above normal (up to 0.8 ppm, 24-hour average). The presence of flu viruses constitutes an additional public health stress agent which may act synergistically with stress due to air pollution. Thus, it is difficult to assign the causes of excess mortality (about 21%) that occurred then solely to air pollution.

Hodgson was the first to analyze a continuous period of monitoring (November 1962–May 1965) using time-series methods [47]. His model analyzed the sum of heart disease and respiratory deaths separately from all other causes, by three broad age groups, and he employed smokeshade, SO_2, and the absolute value of daily degree-days as predictors. The last variable, a temperature correction, considers the separate effects of heat in the summer and cold in the winter, and thus serves as a seasonal correction. This model was used with both monthly data (31 consecutive months) and daily data, including various lag effects and nonlinear specifications. The results showed smokeshade to be more important than SO_2, and the net effect (about 12%) on mortality was about the same as the worst years in London (Fig. 6), despite the less polluted air in New York City and the use of a single monitoring station there to characterize the exposure of the entire city. Hodgson also showed that the monthly models showed larger effects on mortality than the daily models, and that including the previous month's pollution and degree-days (the average of month t and month $t-1$ were used to predict mortality in month t) resulted in a slight increase in the regression coefficient (but with a slight loss in explanatory power). This result suggests that the short-term effects of pollution exposure may persist considerably more than a few days. However, when lags were explored with the daily models, only the inclusion of the preceding day made a significant improvement and SO_2 became statistically significant. Although for Hodgson's monthly results, the pollution effects were larger than the temperature effects, the two were about equally important for the daily models. Respiratory mortality had about the same effects in monthly and daily models, but the effects on heart disease were stronger for the daily model. It is important to note that the period studied included the 1963 flu epidemic and that no special means were taken to account for this variable. Since respiratory and heart disease deaths were combined, all of Hodgson's results may be confounded by this factor, which may partially explain the significance of monthly deaths.

Using mg/m^3 as the uints of measure for the pollutants ($10\ COH = 1\ mg/m^3$) and a total mortality basis, Hodgson derived the following relationships:

For monthly mortality (no lag):

$$\text{percent excess mortality} = 54.8\ \text{smoke} + 2.22\ SO_2 \qquad (17)$$
$$R^2 = 0.73 \qquad\qquad (17.0) \qquad\qquad (4.6)$$

For daily mortality (including previous day):

$$\text{percent excess mortality} = 31.6\ \text{smoke} + 3.2\ SO_2 \qquad (18)$$
$$R^2 = 0.37 \qquad\qquad (4.0) \qquad\qquad (1.16)$$

On the basis of mean values of $0.43\ mg/m^3$ for SO_2 and 0.215 for smoke, Eq. (18)

yields an estimate of about 8.2% for the effect of daily variations in air pollution on New York City mortality, from 1962–5.

Glasser and Greenburg [52] studied the winters of 1960–4 (October–March), including hourly SO_2, bi-hourly smokeshade, daily mean temperature, sky cover, wind speed, and rainfall as potential predictor variables. Mortality data were treated as deviations from 15-day moving averages and deviations from expected values based on the five-year record. Day of week effects were also accounted for. However, in their regression analysis, Glasser and Greenburg treated SO_2 as a surrogate for all air pollution; the result was about 2.8% of mortality associated with SO_2. Their dose-response curve for SO_2 (as the sole pollutant) had an overall slope of about 5 percent excess mortality per mg/m^3, with the suggestion of a threshold at about 0.4 mg/m^3 (24-hour average). With smoke as the sole pollutant, the coefficient was about 15% per mg/m^3, with the suggestion of a threshold at about 3.0 COH. Like Hodgson, they ignored the contributions of the 1963 flu epidemic, so that these findings may well be overestimates. The two-way contingency table they presented giving the deviations in daily mortality for fixed levels of SO_2 and smoke allows some independent estimates to be made of the dose-response relationships. Since each table entry (31 in all) represents the aggregate of a different number of days, different weights will be given to the various pollution levels than in the original data. When all the table entries were regressed on the midpoints of the SO_2 and smokeshade ranges, the following model resulted:

$$\text{percent excess mortality} = 3.6 \text{ smoke} + 7.1 \ SO_2 \qquad (19)$$
$$R^2 = 0.57 \qquad\qquad (3.4) \qquad\quad (1.3)$$

Based on the means of the smokeshade and SO_2 values in the paper, this relationship yields a total mortality effect of 5.0%. This figure is lower than Hodgson's result and the roles of SO_2 and smokeshade as principal contributors have been reversed.

Studies by Buechley et al.

This same period (1962–6) was also studied by Buechley et al. [53], who used data on mortality time trends for 422 urban places in the United States to develop a normalization curve to serve as a daily mortality predictor. For the New York area, they used mortality data from a much larger region, the tri-state air quality control region, which had a population of almost 14 million in 1960. Exposures to SO_2 and smokeshade were based on the single monitoring station at 121st St. in Manhattan. The model accounted for days of the week, holidays, temperature fluctuations, and flu epidemics (important for 1963). Again, the results suggested that SO_2 played a more important role than smoke; the regression coefficient for SO_2 was 4.5 percent excess mortality per mg/m^3. However, the mean value of SO_2 (0.27 mg/m^3) appears to be about half the value expected from other data sources for New York City [52]; using this alternative SO_2 value, the total effect (regression coefficient × mean value) on metropolitan area mortality would be 1.2 percent. If we arbitrarily scale the SO_2 regression coefficient to the "correct" value of mean SO_2 concentration, the SO_2 effect on mortality might have been as high as 2.4%, which is still much lower than found by either Hodgson [47] or Glasser and Greenburg [52].

Buechley subsequently [54] gave more details of his methods. For example, he described calibration problems with the SO_2 instrument in 1964–5, apparently resulting in high readings. An adjustment procedure was made to set the minimum daily values to zero; this may explain the fact that his mean value for 1962–6 is substantially lower than others have cited [52]. Such an adjustment may not substantially affect the resulting regression coefficient, but it will tend to lower the mean value and hence the mortality "effect" attributed to SO_2. Because of the skewed distribution of the SO_2 variable, Buechley also investigated a transformed "Z-score" SO_2 variable which was detrended to remove the effect of the declining mean values from 1967–72 (factor of 5 decrease). He also included a "particulate" variable, assumed to be COH, and a carbon monoxide variable. He stated that CO was missing from 1962–5: it is not clear whether the regression results given for 1962–6 including CO are, in reality, only for 1966 or if some procedure was used to supply data for the missing four years. Buechley [54] gives results for 27 different stepwise regressions in the Appendix to his report, including results for 1967–72, 1962–6, and 1962–72. The temperature, seasonal, and other correcting variables vary somewhat according to the stepwise selection procedure, since typically only about half the possible variables are selected for the optimum step. Buechley points out that since particulates have such a strong seasonal cycle, it is important to use an adequate representation for the seasonal mortality trend in the regressions, and that a simple sine function is not adequate. His regressions include one or the other SO_2 variables and usually particulates, but CO only if the significance criteria for selection is met. The detrended SO_2 variable (SO_2Z) should be regarded as an index variable for all pollutants with the same daily variability as SO_2. For example, SO_2Z is more highly correlated with particulates and CO than is the untransformed SO_2 variable.

Table 2 summarizes Buechley's results for mortality in the New York metropolitan area, and shows that SO_2Z is always more significant than the untransformed variable (however, the mortality effect cannot be computed since the mean Z-score is zero), but this substitution reduces the effects assigned to the other pollutants. The coefficients for SO_2 behave as expected, decreasing with time as the mean SO_2

Table 2. Regression Results for Mortality in the NYC Metropolitan Area; Buechley (1975) [54]

Period	Coefficients (t values)				Percent mortality effect			
	SO_2	SO_2Z	Part.	CO	SO_2	Part.	CO	Total
1962–6	3.5 (5.4)	—	—	—	0.95	—	—	0.95
	—	1.05 (6.3)	—	—	—	—	—	—
	2.7 (4.1)	—	6.2 (2.8)	—	0.73	1.33	—	2.05
	—	0.85 (5.0)	5.5 (2.5)	—	—	—	—	—
1967–72	2.9 (2.1)	—	8.2 (3.3)	0.2 (1.9)	0.16	1.75	0.66	2.57
	—	0.85 (5.0)	6.4 (2.7)	—	—	—	—	—
	0.6 (0.4)	—	9.0 (3.7)	0.2 (1.6)	0.03	1.93	0.57	2.53
	—	0.53 (3.0)	7.5 (3.1)	—	—	—	—	—
1962–72	1.72 (2.9)	—	5.7 (3.8)	0.3 (3.9)	0.26	1.23	1.14	2.63
	—	0.53 (4.4)	4.6 (2.9)	0.3 (3.0)	—	—	—	—

value dropped. The particulate coefficient is almost constant for all the regressions. CO is only significant for the entire period, and since this includes a large block of missing data, these results should be interpreted cautiously. The total effect of air pollution on mortality in the metropolitan area is estimated to be about 2.6%, even after the large reduction in mean value of SO_2. If valid CO data had been available for the earlier period, the total effect then may have been around 3%. If a larger value of SO_2 were used for the earlier period, consistent with other studies [47, 55], the SO_2 effect would approach 1.5% and the total, 4%.

Studies of New York City Mortality by Schimmel and Colleagues

Schimmel and Greenburg [55] studied the period 1963–8, using SO_2 and smoke-shade as predictors and a correction procedure to account for trend and weather effects. Regressions were performed for several different causes of death and for lags up to seven days. A subset of the city immediately surrounding the monitoring station, comprising about 17% of the population, was also analyzed. Schimmel, Murawski and Gutfield [56] and Schimmel and Murawski [57] extended these basic methods to 1972. Schimmel [58] then extended the work to 1976, using slightly different methods, which were intended to supersede those in the previous papers.

The basis for Schimmel and Greenburg's technique was to "correct" the pollution variables for the effects of interfering (collinear) variables, leaving the mortality variables as measured. While Buechley also stated that the dependent variable should remain "uncorrected", he (and Hodgson) used several additional variables in their regressions to deal with various cyclical factors. Schimmel and Greenburg stated that "correcting" the pollution variables for temperature gave almost the same result as adding a temperature variable to the regression. They presented results for uncorrected variables and two methods of correction, which varied by about a factor of 2–3 in terms of the number of excess deaths attributed to air pollution. They recommended the intermediate results as the best estimates, which were, for total mortality, city wide:

$$\text{percent excess deaths} = 29 \text{ smoke} + 3.3 \text{ } SO_2 \text{ (no lag)} \tag{20}$$

$$\text{percent excess deaths} = 46 \text{ smoke} + 4.7 \text{ } SO_2 \text{ (7 day lag).} \tag{21}$$

For the smaller "special district":

$$\text{percent excess deaths} = 26 \text{ smoke} + 6.6 \text{ } SO_2 \text{ (no lag)} \tag{22}$$

$$\text{percent excess deaths} = 49 \text{ smoke} + 11.2 \text{ } SO_2 \text{ (7 day lag)} \tag{23}$$

The regression coefficients for smoke appeared more sensitive to the various correction methods than those for SO_2. Excess deaths due to heart disease and total mortality were both split, 20% associated with SO_2 and 80% with smoke, while respiratory deaths were only associated with smoke. Thus, although no special attention was given to influenza epidemics (1963), there is no evidence of interaction between respiratory deaths and SO_2. For 1963–8 (mean SO_2

$= 0.45 \text{ mg/m}^3$, mean smoke $= 2.1$ COH), the total pollution effects on mortality were estimated to be:

	City wide	Special district
no lag	7.6%	8.5%
7-day lag	11.6%	15.3%

The values for the whole city are similar to those obtained by Hodgson [47], but much higher than Buechley's [54]. The higher results for the special district may reflect errors in measurement of the "true" population exposure for the whole city when a single monitoring station is used.

Schimmel, Murawski, and Gutfield [56] presented results using both uncorrected and "temperature corrected" SO_2 and smokeshade for 1963–8, 1969–72, 1967–9, 1970–2, and 1963–72. Seasonal regressions are also presented, as well as a table of the total pollution effect by year. During this time span (1963–72), SO_2 dropped by about a factor of seven, while smoke remained almost constant. The percentage of excess deaths attributed to air pollution for each year varied from 5.7 to 12.1, independently of the changes in SO_2 concentration from year to year. The year to year variations in the regression coefficients appeared to exceed the year to year variations in mean pollution levels. By season (bimonthly period), the total mortality effects were similar in all periods but July–August, when they were nil. The results for SO_2 suggested behavior as a surrogate, in that as the mean SO_2 value dropped, the regression coefficients, increased giving the same or even an increased effect on mortality. This was the case after 1970; the SO_2 contribution for 1967–9 was markedly lower than for 1963–8. The results for 1963–8 may have been influenced by the 1963 flu epidemic.

The next publication in this series, by Schimmel and Murawski [57], worked with the same data base but used deviations from seasonal trend lines based on 15-day moving averages for temperature, SO_2, and smokeshade. Results were presented for five different mortality measures (by cause). Corrections for day of week and holidays were neglected, but a special effort was made to account for heat wave mortality. The absolute values of percentages of mortality assigned to pollution decreased relative to those in previous publications, but the relative features stayed the same: smoke was more important than SO_2 and there was no decrease in the mortality effect corresponding to the large reduction in ambient SO_2 that had been accomplished.

In the final paper, Schimmel [58] essentially disavowed the early findings that recommended use of the intermediate levels of excess mortality attributed to air pollution, in favor of the lowest estimates. He gave a useful table (his Table 1) explaining the differences in the various regression methods and their results. He also extended the data base to 1976 in this last effort, and results were presented for various lags, by cause of death, and by sex, race, and age group (in Appendices). The following relationships summarize these findings, based on the percentage of (total) mortality attributed to either SO_2 or smokeshade. (The figures in parentheses are the estimated standard errors of the coefficients.)

1. 1963–76, deviation mortality versus deviation temperature and deviation pollution (total effect = 1.59%):

$$\text{percent excess mortality} = 0.74 \text{ SO}_2 + 7.1 \text{ smoke} \qquad (24)$$
$$\qquad\qquad\qquad (1.0) \qquad\quad (2.1)$$

2. 1963–76, deviation mortality versus deviation temperature and deviation pollution, all deviations corrected for temperature (total effect = 0.59%):

$$\text{percent excess mortality} = -1.29 \text{ SO}_2 + 4.8 \text{ smoke} \qquad (25)$$
$$\qquad\qquad\qquad (0.96) \qquad\quad (2.1)$$

3. 1963–9, deviation mortality versus deviation temperature and deviation pollution, all deviations corrected for temperature (total effect = 1.00%):

$$\text{percent excess mortality} = -0.54 \text{ SO}_2 + 5.7 \text{ smoke} \qquad (26)$$
$$\qquad\qquad\qquad (1.00) \qquad\quad (2.8)$$

4. 1970–6, deviation mortality versus deviation temperature and deviation pollution, all deviations corrected for temperature (total effect = 0.26%):

$$\text{percent excess mortality} = -3.9 \text{ SO}_2 + 3.9 \text{ smoke} \qquad (27)$$
$$\qquad\qquad\qquad (2.3) \qquad\quad (3.0)$$

These relationships are based on joint regressions of SO_2 and smokeshade; when separate regressions were performed, SO_2 remained negative but the smokeshade coefficients decreased slightly. The smoke coefficient for 1970–6 was not significant, which could indicate a threshold effect somewhere between 1.77 and 2.15 COHs (mean values for the two periods).

The lag analysis for total mortality showed no changes in significance for SO_2 but the effect of smoke increased with lag time up to a lag of 7 days (the cumulative effect of pollution for the week before death exceeded that for the day of death alone). The effect was in the range 3–4% excess mortality for all three periods. For longer lags, there were negative effects, part of which could be compensation for the "harvesting" of the terminally ill. When the sum of pertinent causes of death was used instead of total mortality, the smoke effects increased by 50–100%, for lags from zero to about four days. For longer lags, the mortality effects for specific causes decreased relative to total mortality, which places further doubt on the reality of the increased effect for lags beyond about four days.

Taken in its entirety, Schimmel et al.'s analysis of New York City mortality is a mathematical tour de force. However, it is difficult to accept Schimmel's final recommendations on statistical methods as given, for several reasons:

1. Measures of the overall goodness of fit often were not given.
2. Statistical comparisons of the various models were not made.
3. Criteria for "satisfactory" or "superior" results were never stated.
4. Treating SO_2 as an index variable by removing the long-term trend confounds its effects with those of other pollutants.

5. The larger percentage effects found in the "special district" were dropped in the later analyses.
6. There was no consideration of the quality of the air pollution measures, and large outliers were truncated, regardless of possible causes.

Other Studies of New York City Mortality

Özkaynak and Spengler [59] and Özkaynak et al. [60] reanalyzed Schimmel's 1963–76 data set and added data on atmospheric visibility, obtained from the three airports in the metropolitan area. Visibility can be a surrogate for the presence of fine particles in the atmosphere, different from the particles which create the dark stain used in the smokeshade measurement. They used a simple sinusoidal model to correct the data for seasonal trend, and restricted their analysis to days for which the visibility records were similar for the three airports. They did not account for days of the week, holidays, or flu epidemics. This regression assigned 1.5% excess deaths to smokeshade, 1.3% to SO_2, and 1.0% to visibility (all statistically significant). They made several other conclusions:

1. Excess mortality attributed to air pollution varied from 2–4%, depending on the methods used for pre-processing the data.
2. Based on analysis of quarterly data, they estimated 2% excess deaths associated with SO_2 and smoke for 1963–9 and 1.4% for 1970–4.
3. The association between mortality and airport visibility was sensitive to the selection of the airport.

Bloomfield et al. [61] studied the effects on time-series mortality regressions of using multiple air quality stations and of disaggregating the metropolitan area mortality counts according to monitoring stations. In the first instance, for coefficient of haze (COH) in New York City from 1971–5 (a period of relatively low pollution), they found that the average of two monitoring stations performed better (gave increased statistical significance) than the average of seven and about the same as the best of the seven stations. Thus, there was no great advantage to either averaging the stations or considering them individually. The effects of SO_2 were negative and nonsignificant; for smokeshade, the effect was in the range 1–2% excess mortality.

In the second part of Bloomfield et al.'s study, three areas of Pittsburgh were considered individually and jointly. Statistical significance was only found for one area, which provided useful information. This study appears to have confirmed that attempts at disaggregation will inevitably suffer from the data "noise" introduced by considering smaller populations. Bloomfield et al. defined "representative" air monitoring stations on the basis of their statistical relationships to daily mortality, rather than on some *a priori* criterion with respect to population locations, a point which detracts substantially from the validity of their analysis. On this basis, they concluded that the 121st St. Laboratory station, which was used for most of the New York City studies, was not "representative" for the 1971–5 period.

Summary and Conclusions from the New York Studies

While most of the studies reviewed here suffered from some defects in data or analysis, it seems safe to conclude that they confirm the finding from the London studies that smoke is a more important indicator for acute mortality than SO_2. This is an important result, considering the different chemical nature of smoke in New York as compared to London, and the much lower concentrations in New York. Table 3 compares selected findings from these studies, based on same-day mortality and the inclusion of no other pollutants in the calculation of "total effect". The regression results fall into two groups: Hodgson and the Schimmel and Greenburg study found large coefficients for smoke and hence large total effects. Neither of these studies used deviations from 15-day averages nor corrected for days of the week, flu epidemics, or holidays. Schimmel and Greenburg accounted for these factors in some of their results, and the smoke coefficients were greatly reduced. In his latest work, Schimmel [58] recommended using these lower coefficient values, but unfortunately he did not give the results for the "special district" on this additional basis. The fact that the total mortality effect did not change much as SO_2 was reduced by a factor of 5 tends to support the finding of smaller coefficients for SO_2 (compared to smoke), although we have no information on the actual city-wide change in SO_2 exposure during this period.

Pickles [62] analyzed the effect of errors in estimating pollution exposure. Based on Goldstein's correlation analysis of the 40-station network [48], he concluded that the use of a single station in New York could bias the results low by about a factor of 2, relative to the seven stations used in London. While section B of Table 3 does not suggest differences that large, it does imply that larger effects will be found for smaller geographic areas, presumably because of better characterization of

Table 3. Results of Time-Series Studies of New York City Mortality

Author [Ref.]	Period	Regr. coefficients		Total effect (%)
		Smokeshade	SO_2	
A. Regressions for the entire city				
Hodgson [47]	1962–5	32–55	2.2–3.2	8.2
Glasser [50]	1960–4	—	5	2.8
Buechley [53]*	1962–6	—	4.5	1.2–2.4
Buechley [54]*	1962–6	6.2	2.7	2.05
	1967–72	9.0	0.6	1.96
Schimmel [55]	1963–8	29	3.3	7.6
Schimmel [58]	1963–9	5.7	− 0.5	1.0
	1970–6	3.9	− 3.9	0.3
Özkaynak [60]	1963–76	7.7	4.8	2.8
B. Comparisons by geographic area (1962–8)				
Buechley [54]	tri-state area	6.2	2.7	2.05
Schimmel [55]	NY City	29	3.3	7.6
Schimmel [55]	special district	26	6.6	8.5

* Tri-state metropolitan area

exposure. This finding also reinforces the causal nature of the air pollution results, since surrogate effects would not be expected to show this trend.

Other aspects of the New York studies suggested that other pollutants, such as CO and fine particles, may affect mortality and that lag effects tend to increase the values of the coefficients and their effects. Most studies showed a lag effect of only a few days.

Time-Series Mortality Studies in Other Cities

Philadelphia

Wyzga [63, 64] studied a 10-year record of data from Philadelphia (the city, as opposed to the metropolitan area), classified by summers and winters (6-month periods) for 1957–60, 1961–3, and 1964–6. For the two earlier periods, only particulate measurements were available (TSP and COHs). Data were available from either two or three sampling stations, depending on the period, which were averaged for use in the mortality regressions. Missing data from individual stations were estimated and added to the record. For the last period (1964–6), data were also available from one station for SO_2, CO, NO_2, NO, hydrocarbons, and oxidants. Wyzga's model accounted for serial correlation, seasonality, heat waves, epidemics, and daily variations in temperature. Lags were investigated for a limited set of data, which showed that only the day preceding death made a significant additional contribution to the same-day regressions.

Wyzga found that COH was more significant than TSP, and thus used COH for studies of the six different periods, with the following results (coefficients in % per mg/m^3 for same-day regressions):

Period	Mean COH	No. of sites	Coeff.	t	Effect (%)
Winters					
1957–60	1.89	2	14.6	2.00	2.8
1961–3	1.61	3	18.8	2.02	3.0
1964–6	1.31	2	35.0	2.85	4.6
Summers					
1957–60	1.22	2	48.8	3.53	6.0
1961–3	0.92	3	33.8	1.82	3.1
1964–6	0.87	2	8.6	0.41	0.8

When lags were added, the winter 1964–6 coefficient increased by about 50%. These data suggest that a threshold of no effect may exist around a mean COH value of about 1.0 (ca. 100 $\mu g/m^3$), and that air pollution in summer may be just as lethal as in winter (at the same concentration level).

The following results were obtained for the winters of 1964–6 for other pollutants (means in $\mu g/m^3$; coefficients in percent excess mortality per mg/m^3):

Species	Mean	Coeff.	t	Effect (%)	r_{COH}
COH	131	35	2.85	4.6	1.0
TSP	162	17	1.85	2.8	0.80
NO	78	28	3.11	2.2	0.81
NO_2	64	20	1.28	1.3	0.60
SO_2	251	3.5	0.94	0.9	0.67
CO	8640	0.24	1.35	2.1	0.33
HC	1483	3.1	2.17	4.6	0.69
Oxidants	40	16	0.53	0.6	0.40

The last column in the table gives the correlation between each species and COH, for 1964–6, which may be used to speculate whether the effects shown for each species relate to the health effects of that species or just to a statistical artifact arising because of correlated measurements (not to be confused with correlated *exposures*). Of the pollutants above, only COH, NO_2 and possibly SO_2 may be assumed to be widely distributed throughout a metropolitan area. CO, NO, and hydrocarbons (HC) are primary pollutants from traffic and would not be expected to persist in non-traffic areas. Oxidant levels (primarily ozone) tend to be depressed in traffic areas because of chemical reactions with NO, and thus levels may be much higher in the suburbs. Therefore, the statistical significance of HC may be an artifact (especially since most of this species is methane, which is thought to be relatively innocuous for health effects). The result for NO is more difficult to interpret, since there are other sources of nitric oxide emissions besides traffic (all stationary combustion sources) and this species was more statistically significant than COH. NO could conceivably represent a surrogate for nitric acid (but NO_2 would probably be a more realistic surrogate). Analysis of other cities, seasons, or time periods is required before effects on mortality of any of the other species can be considered more seriously.

Pittsburgh

Mazumdar and Sussman [65] continued the most recent methodology developed by the Schimmel group, for both mortality and morbidity in Allegheny County, PA (Pittsburgh metropolitan area) for 1972–7. They used deviations from 15-day moving averages, corrected for temperature by a two-stage regression procedure, with smoke and SO_2 measured at three different stations. Dummy variables were used to account for day-of-week effects. Data from each pollution measurement station were entered separately in the regression; the correlations between stations were all less than 0.6. In general, only the most polluted station showed statistical significance, for smoke in a joint regression with SO_2 (which was negative). None of the pollutants was significant in separate regressions. The smoke coefficient for total mortality was about 8% per mg/m^3, for the station with a mean COH of 1.27 ($SO_2 = 0.046$ ppm). When mortality for the local area around the monitoring station was regressed against the local smoke values [61], the smoke coefficient increased to about 12% per mg/m^3, which is consistent with many of the previous studies. The finding of nonsignificant SO_2 effects is not surprising, given the modest concentration levels present (mean values from 65 to 123 $\mu g/m^3$). Note that the period studied includes the 1975 "episode" in Pittsburgh, discussed above, during which TSP reached 770 $\mu g/m^3$ and SO_2, 200 $\mu g/m^3$. Regressions for heart disease mortality provided higher coefficients, but they were no more statistically significant than for total mortality.

Los Angeles

The pollutant species of interest in Southern California has traditionally been ozone or total oxidants. Mills [66] performed a time-series analysis of mortality in Los Angeles County due to cardiac or respiratory causes, for 1956–8. Daily maximum oxidant readings from up to 20 stations were used to provide a county-wide average. Daily maximum temperature was used as a partial control: days with maximum readings over 96°F were omitted from the trend analysis, which was based on deviations from monthly averages, on two bases: all available days (881 days) grouped according to average maximum oxidant reading, and the same groupings for the months of January through May, only. Both methods provided statistically significant slopes through the grouped data corresponding to about 23 percent per pphm (Fig. 17). There is no suggestion of a threshold effect, even at levels corresponding to the current ambient air quality standard (12 pphm), but there may be a slight suggestion of saturation above 32 pphm. Using the mean of Mills' tabulated oxidant data, the average mortality effect in Los Angeles County was about 3.2% year-round and 2.8% in the winter-spring period. These percentages would probably be somewhat lower if expressed on a total mortality basis.

Since Mills tabulated the entire data set, it is possible to get a feeling for the interrelationships by inspecting the data. Heat-wave mortality was apparent, and days with repeated high maximum oxidant readings appeared to be especially lethal. However, there were also high mortality days not coinciding to either

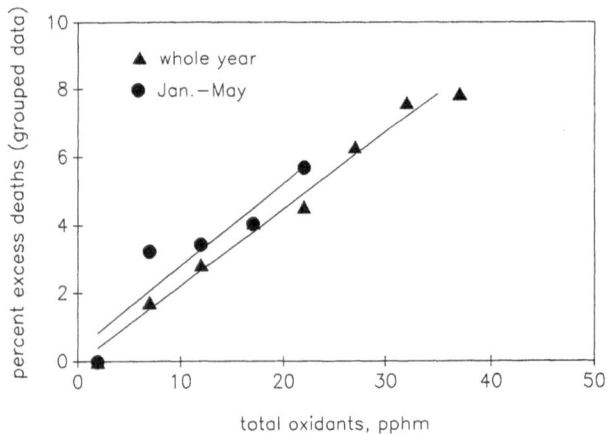

Fig. 17. Dose-response functions for excess mortality in Los Angeles as a function of oxidants, based on deviations from monthly averages. Data from Mills (1960) [66]

temperature or oxidant peaks. Joint regression of the monthly average deaths on monthly average maximum oxidants and temperature yielded a non-significant result for oxidants (in contrast to Hodgson's [47] findings for SO_2 in New York), suggesting that the oxidant episodes had minimal lasting effects. Occasionally, there were dips in mortality following the oxidant or temperature peaks, suggesting a "harvesting" effect.

Hexter and Goldsmith [67] studied total mortality and mortality from heart disease in Los Angeles County for 1962–5 with respect to "basin averages" for carbon monoxide and oxidants (number of stations not given), in conjunction with cyclic trend and temperature variables. Although the logarithm of CO was statistically significant, the degree of fit for this model was only marginally better than for a model with all other terms except log CO. The total mortality effect was 13.5%, substantially higher than Wyzga found in Philadelphia from a single CO station at about the same mean value, and also substantially higher than Mills' result for oxidants in L.A. If Hexter and Goldsmith's CO coefficient is converted to linear form (as an approximation), the result is 0.44% per mg/m^3, which is not greatly different from Wyzga's value for Philadelphia. It is difficult to compare their results with Buechley's result for CO from joint regressions in New York because of uncertainty regarding the units of measure. However, the mortality effect attributed to CO there is small (1%), in part because of the use of a single monitoring station. Hexter and Goldsmith did not report their results for oxidants, only that the regression errors were larger when this pollutant was used.

Chicago

A time-series analysis of daily deaths in Chicago is discussed on p. 81, in conjunction with a larger study program emphasizing cross-sectional analysis. The two types of studies are compared in that section.

Athens

Hatzakis, et al. [68] analyzed daily mortality in the Greater Athens area for 1975–82. The dependent variable was the deviation from the expected daily mortality count, i.e., "adjusted" mortality. Expected mortality was based on fitting a sinusoidal curve to the monthly mortality data for 1956–8. The independent variables were the 24-hour averages of SO_2 (mean = 86 $\mu g/m^3$) and British smoke (mean = 57 $\mu g/m^3$), calculated as the geometric means of data from five measuring stations. During the eight years studied, air quality (maximum monthly average) improved by about a factor of five.

Controls (dummy variables) for day-of-week, month, season, year, holidays, and temperature were included in the analysis. The authors reported that adjusted daily mortality was significantly associated with SO_2 (p = 0.05), but not with smoke. If a threshold were present for SO_2, it lay slightly below 150 $\mu g/m^3$. No regression results were reported for smoke or for SO_2 and smoke taken jointly, although one of the authors stated in a personal communication [69] that the p-value for smoke alone was 0.67 and that they felt that this lack of significance resulted from the greater spatial variability of smoke relative to SO_2.

This study is important since it represents a finding of daily mortality effects at or below the current U.S. primary standard for ambient SO_2. The analysis may have been compromised by evaluating only one pollutant at a time, which the authors felt to be necessary because of the high inter-correlation (0.73). The regression coefficient reported is larger than any reported for either SO_2 or smoke in several analyses of daily mortality in New York and London, although the SO_2 result just met the significance criterion ($p = 0.05$). A stepwise regression approach with both pollutants initially in the regression would be particularly interesting, since so many potential confounding variables were investigated. There are also possible contributions from other (unmeasured) air pollutants; Hatzakis et al. reported a mean of 158 $\mu g/m^3$ for NO_2, measured from 8 A.M. to 2 P.M. on 834 days [69], which is high by U.S. standards.

Osaka

Watanabe and Kaneko [70] briefly reported the results of a five-year (1962–7) correlation study in Osaka, Japan for SO_2, suspended matter, and temperature. Although they presented no multiple regression results, they concluded that SO_2 had the greater effect, at cold temperatures. SO_2 levels occasionally exceeded 0.2 ppm and suspended matter, 1.0 mg/m^3. No data were given on the magnitude of the effect nor on regression coefficients.

Dublin

Kevany et al. [71] used a partial correlation analysis on mortality and hospital admissions data (by cause) in Dublin for the winters of 1970–3. An eight-station monitoring network for smoke and SO_2 was used; two other stations had been rejected because of the influences of local sources. Maximum daily temperature was used to control for weather effects; the study was limited to October–March. The statistical technique was that of partial correlations of selected variables, controlling for the others (regression coefficients were not given). The data set was truncated below increasing pollutant concentration levels to examine threshold effects. Cardiovascular disease mortality was associated with SO_2, only on the same day; smoke was not so associated. The partial correlation coefficients increased with threshold level; for SO_2, in excess of about 75 $\mu g/m^3$. The primary contributing cause of death was ischemic heart disease. At longer lag levels (seven days), smoke above 100 $\mu g/m^3$ showed a strong partial correlation. The authors concluded that ". . . SO_2 has immediate effects on cardiovascular mortality which are compensated for by reductions on the following days. However, smoke showed a sustained increase in mortality over a longer time span . . . ." The mortality results were supported by hospital admissions correlations for both smoke and SO_2. The correlations for respiratory mortality were only significant for SO_2 over 75 $\mu g/m^3$ for acute bronchitis (no lag) and chronic bronchitis (seven day lag). No respiratory causes were significant for smoke.

Vienna

Neuberger et al. [72] studied the period 1972–83 in Vienna, with emphasis on influenza morbidity and mortality among the elderly (> 70). SO_2 was the only pollutant considered, with temperature as a covariate. Both the presence of influenza and SO_2 (probably as an index of all air pollution) influenced total, cardiovascular, and respiratory mortality. For total mortality, the SO_2 coefficient for the over-70 age group was about 36% per mg/m^3 and did not vary substantially between men and woman nor between normal periods and periods of influenza (the overall mortality rates were higher during influenza periods). This finding implies that there is no interaction between air pollution and influenza, and that neglecting influenza in regression analysis should not create a major error, at least for this age group and level of air pollution (SO_2 up to 500 $\mu g/m^3$).

Conclusions from Time-Series Studies

The works reviewed above present a range of methodologies and results. They confirm that air pollution has a significant effect on mortality, even at concentration levels much lower than occurred during the episodes, and, in some cases, lower than present air quality standards. While reliable thresholds for these pollutant effects have not been identified, by and large none of the studies reviewed were structured to find such thresholds. Several studies suggested that thresholds for both SO_2 and smoke may be around 200 $\mu g/m^3$. It is difficult to identify with certainty the most important pollutants in many of the cases reviewed.

The review has shown the importance of accounting for confounding trends and cycles, such as seasons, daily temperature, and days of the week. It is not possible to identify any one method of doing this as "superior" to any other method.

Although most studies which compared SO_2 effects to smoke effects at the higher concentration levels concluded that smoke was the greater hazard, SO_2 cannot be completely exonerated. Figure 18 compares estimates of the total air pollution effect on mortality, plotted against the sum of SO_2 and smoke (a crude but simple way of dealing with joint exposures used here for illustration). The regression lines

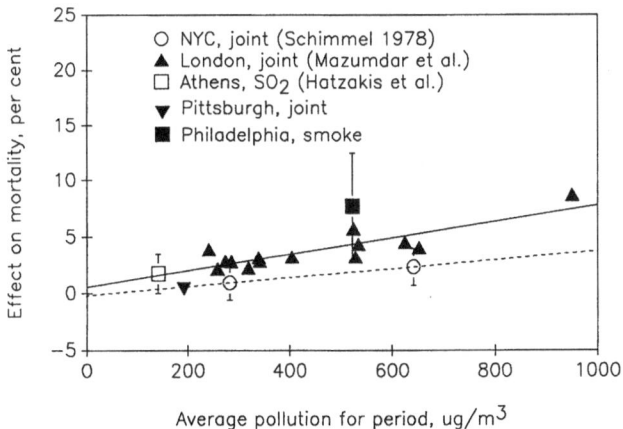

Fig. 18. Comparison of air pollution effects on mortality for various cities. Error bars represent 2-sigma confidence limits. The lower line is for the New York City study; the upper line is a linear regression of all data points. Data are plotted vs. the sum of SO_2 and smoke

pass through the origin, but suggest a smaller effect for Pittsburgh and New York (lower line). The Pittsburgh point would fall nearer the upper regression line if the separate regression for smoke were plotted, using the local area data. The New York regression coefficients may be depressed by the use of a single (and probably non-representative) monitoring station, by as much as a factor of 2 [62]. Given these considerations, the upper line appears to fit most of the data, and may be represented as:

$$\text{percent excess mortality} = 8\,(\text{smoke} + SO_2)\,(\text{mg/m}^3) \qquad (28)$$

This finding is in good agreement with the results of the analyses of isolated episodes presented above (11% excess mortality per mg/m^3 smoke as the sole pollutant), but should not be interpreted as demonstrating that exposures to smoke and SO_2 are of equivalent severity. It is not clear whether the mortality effects of other co-varying pollutants such as CO or oxidants are implicitly included in this relationship. For example, based on Eq. (28), during an event with 24-hour average SO_2 and smoke each at 150 $\mu g/m^3$, about 2.5% excess mortality would occur. This would be about one death per day for a city of two million inhabitants. While this analysis has not been able to define thresholds of "no effect", from a practical perspective such a threshold may result from the statistical inability to detect excess mortality at the lower concentration levels. This situation results, in part, from the large populations needed for such a study and the concomitant difficulty in accurately quantifying their exposure to air pollution.

A potentially important question concerns short-term mortality effects from sustained, multiday episodes. Most studies have looked only at single-day air concentrations, whereas the worst episodes persisted for several days. Only Schimmel [58] and Wyzga [63] considered cumulative effects in their lag analyses, and they did not distinguish the effect of episode duration separately from concentration.

Long-term (Chronic) Effects on Mortality

As stated by Lazar [73], "... an observed variance is at the source of any epidemiological knowledge . . . ." In the short-term studies discussed above, these variances occurred over time. To study long-term effects, it is usually necessary to study variation with respect to place (location).

The study of long-term or chronic health effects of air pollution has been fraught with difficulty and controversy, more so than the short-term studies. In this section, the primary method considered of deducing long-term environmental effects in humans involves comparing the health statistics of populations of places which have had different environments over the long term; such cross-sectional comparisons of environmental differences constitute the third class of "natural experiments". However, the comparisons are often compromised by other differences that may be related to the sources of air pollution, such as industrialization.

Methodological Issues

This type of observational epidemiology, the cross-sectional analysis, has been called "ecologic" regression analysis. It seeks to draw inferences about the outcomes to (grouped) individuals based on the statistics of the behavior of groups (rather than individuals), usually based on very limited information about their exposures to potential causal agents.[1] In addition, the types of group information that are available often do not include important individual lifestyle variables such as smoking, exercise, and diet. Insurance company and survey data have shown that mortality rates can vary by a factor of 2 to 3 among groups differing in lifestyles [74, 75].

The cross-sectional analyses reviewed in this section attempted to define associations between premature mortality and air pollution by analyzing spatial patterns. Typically, annual mortality rates are computed for a set of observational units (political subdivisions), appropriate demographic control variables are defined and data obtained from census sources, and air pollution exposures are estimated from extant monitoring data. Ideally, these data will all be for the same period, unless it is desired to test for a lag between air pollution exposure and subsequent mortality (when pollution levels have varied over time). Multiple regression analysis is then used to develop a "model" for mortality rates; some authors have used such models to infer causal relationships between mortality and air pollution. These studies are usually described as seeking chronic health effects, since no account is taken of daily variations, except as reflected in the annual averages or totals.

Such cross-sectional studies represent one of the few ways that chronic health effects may be studied at reasonable cost in large populations. However, there are problems with respect to the data used and the interpretation of results. The study design requires that the average properties of the population groups be interpreted as appropriate to each of the individuals comprising the group. All of the multiple regression variables refer to properties of the group, yet the group mortality rate stems from individuals who died for various individual reasons. This is often referred to as the "ecologic fallacy" [76]. For example, if a group has both high rates of mortality and high consumption of tobacco (presumably because of an excess of heavy smokers), the ecologic assumption is that the heavy smokers in the group are the ones who died prematurely. This type of group regression analysis does not permit verification of such assumptions, except as inferred from statistical robustness. When responses are nonlinear, such that a (group) average value of an independent variable does not correspond to the average group response, this type of analysis may yield incorrect results.

In addition, it is generally not possible to distinguish between long-term (chronic) effects and the annual sum of acute effects when only one year is analyzed. To estimate long-term effects where the situation has changed over time, historic

---

[1] To the extent that air monitoring data may not accurately represent the actual exposure of affected individuals, time-series studies may also be deemed "ecologic."

values of the independent variables (especially air pollution exposures, i.e., cumulative dose) should be used.

Effects of Errors in Measurement

The effect of error in the measurement of an independent variable is to decrease both its statistical significance and its regression coefficient relative to the "true" values [77], assuming that the effect was in fact real in the first place (in the statistical sense). Since the concentration patterns of many pollutants tend to be collinear, this may mean that the pollutant with the least error in exposure estimation may appear to be the most statistically significant, even when the real situation is that a mix of pollutants is responsible for the observed effect. In such situations, measurement errors can mask the true effects. Of course, none of these considerations has any effect on the issue of causality versus association.

In order for measurement errors to *increase* the significance and regression coefficient for a variable, the error must be a selective bias and not just random noise. An example might be better measuring instruments or characterization of exposure in areas of high mortality rates, compared to areas of lower mortality rates.

Selection of Geographic Units for Cross-Sectional Analysis

The population groups considered in such ecologic studies range from parts of cities to entire states or even countries. (One of the first identifications of the link between cigarette smoking and lung cancer was a cross-sectional regression analysis involving only eight countries [78].) In the United States, many studies have used the "Standard Metropolitan Statistical Area" (SMSA), which is an aggregation of counties surrounding a major city. Since exposure to air pollution may be correlated with other characteristics that can affect longevity, to avoid confounding, cross-sectional regressions should address all relevant attributes that vary spatially. Unfortunately, this is rarely possible. Such variation may often be reduced by using the smallest possible geographic units of analysis. Piantidosi et al. [76] compared regression coefficients for a group of physiological and dietary variables, contrasting the "true" values obtained from individuals with those obtained from various aggregations. In four out of the thirteen cases, the true values lay outside the confidence limits for the aggregated values, with errors in both directions.

The accuracy of estimating exposure to air pollution will vary with the size of the political subdivision used and the nature of the pollutant. Some primary pollutants, such as TSP, CO, and SO_2, tend to be very local, and concentrations may vary substantially within a few city blocks, in addition to varying between indoors and outdoors.

Secondary pollutants, such as NO_2, oxidants, and sulfate particles, may exhibit less spatial variability, although ozone can be strongly attenuated locally by the presence of NO_x sources. Most cross-sectional studies have had to work with data from a few air pollution monitors and have made arbitrary assumptions about the size of the area that each monitor represents. The lack of true representation of the air pollution exposure of the population constitutes an important source of error in the independent variables.

This source of error is also associated with the choice of the type of political subdivision for the observational unit, since the larger its area, the larger the chances for errors in estimating true population exposures (assuming a fixed number of monitors and that local pollution sources are present).

For example, assume that there is a true relationship between particle concentration and mortality (this need not be a causal relationship, since there may be other aspects of the pollution source to consider, such as occupational factors). Often there have been available two measures of particle concentration: total suspended particulate matter (TSP), which tends to be somewhat local because it may include particles up to 50 μm in diameter; and the sulfate portion of the particulate catch, which is usually distributed regionally since the particles are much smaller and travel further. (Recently in the United States, particulate monitoring has separated fine and coarse particles.) When relatively small areas (such as cities or portions of cities) are used as the observational units, TSP exposures may be reasonably well-represented. On the other hand, if larger units are used with the same monitoring network, such as entire counties or metropolitan conurbations, any "true" TSP effect on mortality is likely to be masked by the exposure error, since many of the people "assigned" to the TSP monitor live so far away that they are not actually exposed to the pollution measured there. Now, if at the same time there is a regional trend towards higher mortality in the region of high sulfates (or any other regionally-distributed pollutant), the regional pollutant will become the significant variable. This result may appear to be a health-based causal finding, since small particles can penetrate deeper into the lung, but, in this case, the result appeared as a statistical artifact because a regionally-distributed pollutant was matched with a regionally-distributed mortality trend. An analysis based on large geographic units is unlikely to capture local pollution effects, only regional ones, but a city-based analysis should be able to detect either type. This distinction is similar to separating the high-frequency (short-term) effects from the seasonal effects in a time-series analysis.

However, mortality rates may be statistically unstable if the population base is too small. One solution to this problem is to use small geographic areas (i.e., central cities) with data averaged over several years, which will improve the stability of estimates of both mortality and air pollution exposure.

Confounding by Demographic Factors

Other problems have arisen because of collinearity between air pollution and demographic variables, which is always present to some extent. Collinearity can arise from the fact that pollution sources have other types of impacts upon their host communities, in addition to exposing them to air pollution. Heavily industrialized cities tend to have demographically different populations than, say bedroom suburbs for large metropolitan areas. These population differences typically include education, smoking habits, race and income, all of which can have independent effects upon average mortality rates. Such effects may be less apparent when larger observational units are used, because of averaging. However, if·the pollution exposure is not averaged because the available monitoring data reflect only the air quality in the polluted part of the area, the multiple regression results will vary according to the size of the observational unit. Thus, this source of collinearity is more of a problem in central cities than in SMSAs, but the reduction in collinearity for SMSAs is only apparent, not real, since it results from the use of central city air quality data to (inadequately) represent an entire SMSA.

It is not possible *a priori* to establish acceptable limits on multicollinearity. If effects such as education, smoking, or income are not explicitly represented in the regression models, they may be erroneously (implicitly) represented in part by the air pollution variables. Such problems can only be handled by incorporating the appropriate independent variables in the analysis and then sorting them out by stepwise or other repetitive regression methods. Factor or principal component analysis may be another approach. However, when repetitive specification searches are made, statistical significance levels may be overstated (due to loss of independence).

Population migration patterns also can cause errors in estimated pollution exposures, as well as confounding of regression results. Confounding results from either selective migration of sick people or of the more economically advantaged. In either case, current (local) air quality may not represent the true long-term exposures of current residents.

Other problems can arise when unadjusted total mortality data are used (all causes, ages, races; both sexes). Often, for smaller geographic subdivisions, only this type of data is available. Age adjustment is the most important correction to make, since the probability of dying in a given year increases exponentially with age above about age 30. If mortality rates are available for detailed age groups, they can be combined into one age-adjusted total rate by reference to the age distribution of a standard population. If, on the other hand, only total deaths are available but details are available on the population's age distribution, then the expected total number of deaths may be computed on the same basis. In many cross-sectional studies, neither procedure was followed, but surrogate age adjustments were attempted by using a population age descriptor variable as an independent variable in the multiple regression. "Percentage of population aged 65 and over" is a common choice. If all populations have similar age distributions, such a choice may be acceptable, but simple algebra shows, for example, that the regression coefficient for "% > 65" should be numerically equal to the mortality rate for this age group minus the rate for the under 65 group [80]. Many studies do not meet this simple test. Similar considerations apply to other explanatory variables employing percentages of the population, such as "% nonwhite" or "% poverty".

Regression Models and Causality

Such regression methods often appear to be rather *ad hoc* and have been called "data mining" or "fishing expeditions" [79]. The problem is lack of a theoretical *a priori* "model" (in the econometric sense) for human mortality and morbidity, as well as lack of appropriate data bases for the known risk factors. For example, for heart disease, which accounts for a large fraction of all deaths, the known risk factors include smoking, diet, exercise, lung function, and perhaps drinking water

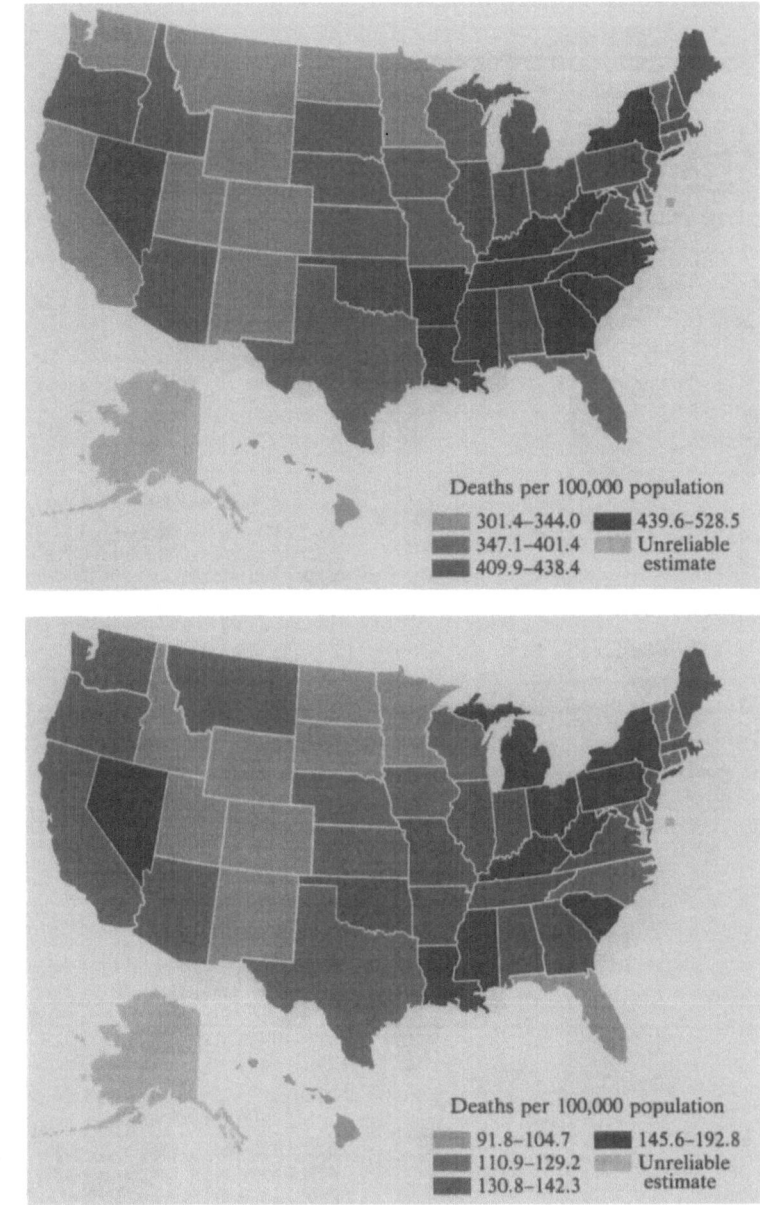

Fig. 19. Heart disease mortality patterns in the U.S. Source: Health, United States, 1988 (U.S. Dept. of Health, Education and Welfare). **a** White males, ages 45–64. **b** White females, ages 45–64 (1983–85 data)

hardness and ethnic background. A few of the well-known cross-sectional studies have attempted to represent some of these risk factors; most have not.

Spatial patterns of U.S. mortality rates show some well-defined trends. Heart disease rates are higher east of the Mississippi and especially in the Southeast

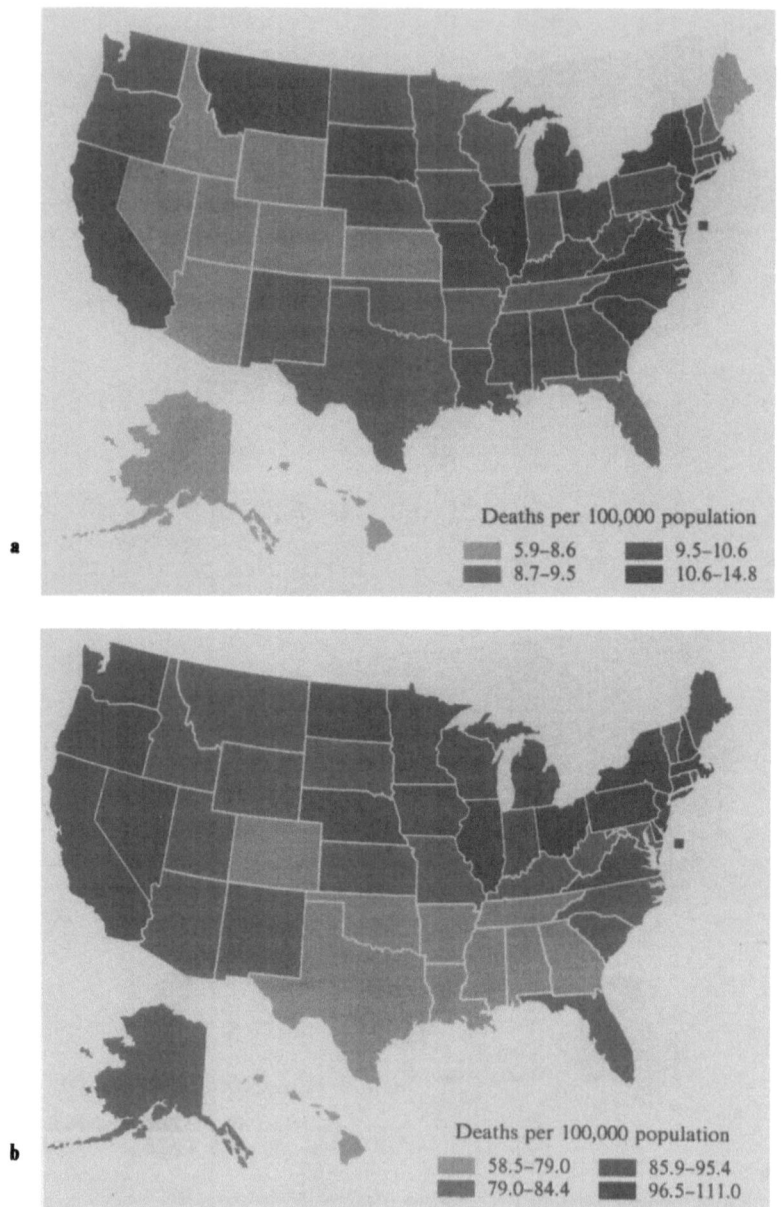

Fig. 20. Breast cancer mortality patterns in the U.S. Source: Health, United States, 1988 (U.S. Dept. of Health, Education and Welfare). **a** Females, ages 25–44. **b** Females ages 55–74 (1979–85 data)

coastal plain (Fig. 19). Cancer rates (as exemplified here by female breast cancer) tend to be higher in the northern industrialized states (Fig. 20). Respiratory death rates are higher in the West. These patterns may suggest associations with certain air pollutants as well as with other variables; the challenge to the epidemiologist is to establish whether the associations are causal or merely circumstantial.

There are no standard tests for causality, and Lave [81] even suggested that causality is only a theoretical construct, based on a consensus of scientific opinion. For long-term health studies, establishing causality requires that the known clinical effects of the pollutant are consistent with the health end-point in question, that associations based on spatial relationships are consistent with those based on temporal relationships, and that all reasonably plausible confounding variables are accounted for.

Feinstein [82] postulates five "scientific standards" for valid epidemiological studies: (1) A stipulated research hypothesis (as opposed to a "fishing expedition"), (2) a well-specified cohort, (3) high-quality data, (4) analysis of attributable actions, and (5) avoidance of detection bias. Although not all of these "standards" apply directly to the study of the effects of community air pollution, the requirements above should be kept in mind reviewing the specific studies which follow.

Early Cross-Sectional Mortality Studies (Data from 1930–60)

Mills and Mills-Porter

One of the first cross-sectional studies of the effects of air pollution on mortality was performed by Mills and Mills-Porter [83], who used dustfall statistics as an index of the spatial air pollution gradients in Chicago, Detroit, Pittsburgh, Cincinnati, and Nashville. Atlanta was grouped into pollution gradients, based on subjective classifications by the chief smoke inspector. The mortality indices used were deaths for white males and females for various years from 1929–46 due to pneumonia, tuberculosis, and respiratory cancer (Chicago and Detroit only). The statistic used to establish a relationship between mortality and air pollution was the chi-squared test for differences between the five dirtiest and the five cleanest areas within a city. These differences were statistically significant for males in all cases, but the effects for females were often substantially smaller. This result could indicate confounding either by smoking (which the authors recognized as a possible synergistic effect) or by occupational exposures. Also, no mention was made of age adjustment, a possible further source of confounding. Comparison of the mortality rates for the "cleanest" areas of each city for the same time periods gave no suggestion of regional effects or effects of city size; the "clean" rates in Nashville were substantially higher than in the other cities.

In a subsequent report, Mills [84] extended the Chicago analysis to include age-specific mortality (no changes in conclusions) and discussed the parallel air quality trends for SO_2, confirming that the dustfall statistics used in the first paper should be regarded as a general index of air pollution.

Correlation Analysis of U.S. SMSAs, 1949–51

Manos and Fisher [85] performed an exploratory investigation of interrelationships between several indices of industrialization and air pollution potential and standardized mortality ratios for 50 causes of death. No actual measurements of air pollution were used, and no specific account was taken of possible confounding by socioeconomic factors or smoking. The seven causes of death most frequently linked to the various pollution indices were esophageal and stomach cancer, coronary heart disease, lung cancer, endocarditis, chronic rheumatic heart disease, diabetes, and oral cancer. For the remaining causes, there was a noticeable shift in the distribution of correlation coefficients towards nonsignificant or negative values. Since most of these diseases are also linked to smoking [86], there is strong suggestive evidence of confounding of these results by differences in smoking habits, stomach cancer being a notable exception. The causes of death potentially associated with air pollution (or smoking) on physiological grounds or by other studies that were *not* well-correlated with these air pollution indices included pneumonia, hypertension, bronchitis, tuberculosis, influenza, emphysema, and asthma. This technique of exploratory analysis was supported by some researchers [87] and criticized by others [88]. The latter paper discusses possible confounding by smoking habits, and presents data on mortality by sex and also on the ratios of mortality in central city counties to non central-city counties, by cause and sex. It also speculates that the role of air pollution in heart disease might be to exacerbate the symptoms of those with already impaired circulatory systems.

The Erie County Air Pollution-Respiratory Function Study

Erie County (Buffalo and environs), New York was studied by Winkelstein and colleagues during the 1960s [89–91]. Pollutants were measured during 1961–3 at 21 sampling stations and included TSP, dustfall, and sulfation rate (a crude index of airborne sulfur compounds).[1] The method of statistical analysis was the two-way contingency table for census tracts grouped by location. Only one pollutant was considered at a time, in conjunction with "economic level" (median family income for the group of census tracts). Data were also given on years of school completed,

---

[1] The correlation between TSP and dustfall was 0.84, and the regression between the two (TSP [$\mu g/m^3$] = 40.7 + 80.6 * Dustfall [mg/cm^2mo]) allows rough estimates to be made of TSP in other locations based on dustfall measurements. Such estimates are useful in comparing environmental conditions during the older studies versus the more recent ones. According to this formula, the following long-term average TSP values would pertain ($\mu g/m^3$):

City	Period	Max TSP	Av'g TSP	Ref.
Buffalo	1961–3	206 (as TSP) 161 (from dustfall)	112	[89]
Chicago	1946 (winter)	539	177	[83]
Detroit	1941–2 (winter)	550	299	[83]
Pittsburgh	1938–40	728	313	[83]
Cincinnati	1930–40	442	144	[83]

percentage of laborers in the labor force, and percentage of sound housing for the groups of census tracts in the contingency tables. Thus, the Erie County study was among the first to recognize the need to consider socioeconomic factors in detail in conjunction with cross-sectional analysis. The study populations were white males and females, aged 50–69 and 70 and over.

In the first two papers [89], the authors examined 1959–61[2] mortality from all causes, chronic respiratory disease, respiratory cancer, and those death certificates with any mention of asthma, bronchitis, or emphysema; deaths related to failures of the circulatory system were analyzed in the third paper [90] and stomach cancer was examined in the fourth [91]. While the results for males may have been confounded by smoking patterns, there was no consistent relationship between the average level of air pollution and lung cancer, which suggests that such confounding was not very serious. The contingency tables show a consistent increase in mortality with TSP for all the causes of death analyzed; the trends persisted when mortality ratios were standardized for economic levels and plotted against air pollution level. On this basis, the effect of TSP appeared stronger for females than for males (all causes of death and heart disease), which could imply minimal confounding by smoking or by occupational exposure. In contrast, the results for sulfur oxides appeared less consistent (as published), and when both pollutants were considered together, the effect seemed to be entirely due to particulates. However, the correlation between the two species was very high (0.98), so that it seems futile to attempt to separate the effects.

The analysis of arteriosclerotic heart disease (ASHD) and cerebrovascular (stroke) mortality [90] considered only TSP and showed strong effects for both males and females, only in the 50–69 age group. For ASHD, the effects of economic grouping were strongest for the lowest economic group, and the air pollution effect was ambivalent within the highest economic group.

The stomach cancer study [91] was conducted to investigate the possible role of ethnicity in these cross-sectional studies because certain ethnic groups were thought to be more susceptible to stomach cancer, presumably because of their diet. The authors concluded that stomach cancer was positively associated with TSP for both males and females, aged 50–69, independent of economic level and ethnic origin. They cited support for this finding from previous studies by Stocks in Britain [92]; Manos and Fisher also identified an association between indices of air pollution and stomach cancer in the United States [85]. Presumably, the causal mechanism involves swallowing carcinogenic particles that have been cleared from the respiratory tract.

One of the important findings of the Winkelstein studies is the interaction between economic level and air pollution with respect to residence; in Buffalo, the contingency tables implied that persons of higher economic status tended to avoid the more polluted residential areas. Based on the tables, this was more the case for TSP than SO_x; however, the correlation matrix in Table 4 (below) shows the same effect for both pollutants. This correlation between independent variables can lead

---

[2] The authors assumed stationary patterns in mortality and air pollution, since the time periods did not coincide.

to errors in the contingency table analysis (as discussed in the original paper [89]); thus, for the purposes of this chapter, the Erie County data were reanalyzed using multiple regression methods.

The purpose of this reanalysis was to confirm the authors' conclusions, which had been based on analysis of the 2-way contingency tables and standardized mortality ratios. Some of the results are shown in Figs. 21 to 26, which are plotted

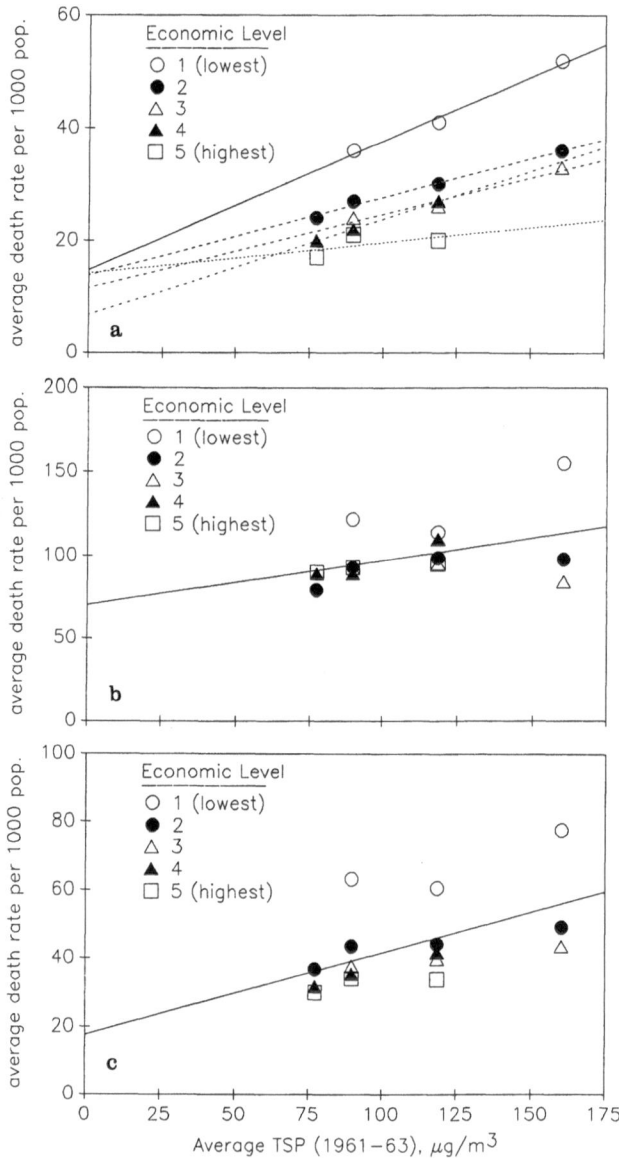

Fig. 21a, b. All-cause mortality for white males in Erie County, NY, 1959–61, by average TSP level. **a** Ages 50–69. **b** Ages 70 and over. **c** Ages 50 and over. Data from Winkelstein et al. (1967) [89]

against TSP as a matter of convenience. Since TSP and SO_2 are so highly correlated in this data set, either pollutant could have been used for the graphs. Figure 21a shows the all-cause mortality data for white males, age 50–69. The interaction between economic level and TSP is readily apparent, since the individual regression slopes against TSP for each economic group become weaker as economic level rises. Thus, a unit of TSP appears to have a larger effect on a male of lower economic status. This effect, if real, highlights the importance of obtaining detailed data on air pollution exposure in conjunction with socioeconomic factors, rather than using city-wide averages (in a multi-city study).

Figure 21b presents the all-cause mortality data for males aged 70 and over. The TSP effect is not statistically significant, and the economic effect seems confined to the lowest economic group; these results may be due to the lower range of variation among census tracts for this age group. Combining the two age groups (Fig. 21c) provides a closer comparison to the crude (total) mortality used in many other studies; the economic effect again appears confined to the lower group, the overall TSP effect is significant (p = 0.05), and the interaction is weaker than in Fig. 21. For comparison, the 1960 U.S. annual average mortality rates for white males (all causes) are 22.5 per thousand population (age 50–69), 38.6 (age 50 +), and 77.8 (age 70 +) [93]. These figures imply an effective TSP intercept or threshold of around 50–75 $\mu g/m^3$.

Figures 22 and 23 present circulatory system mortality for males and females. Figure 22 shows the strong dependence of ASHD mortality for both males and females in the 50–69 age group (Figs. 22a and 22b); the slopes of the overall regression lines are about the same, even though the average mortality rate for females is about half that for males. Note that the air pollution effect is inconsistent for both sexes in the highest economic level group. For the 70 + age group, both air pollution and economic effects are inconsistent (Figs. 22c and 22d). Cerebrovascular mortality rates are displayed in Fig. 23; males and females are plotted together since the gender differences were minimal. The economic effects tend to be less consistent for this cause of death, especially for females.

Figures 24–26 present mortality data for various respiratory causes of death for white males, age 50–69. Respiratory cancer deaths are plotted in Fig. 24; the economic effect is evident, but the overall TSP effect, although positive, is not significant. Figure 25 presents similar data for deaths due to chronic respiratory disease (underlying cause); the overall TSP effect is highly significant, although not all of the slopes for individual economic strata are consistent (not shown). When mortality with any mention of respiratory disease on the death certificate was analyzed (Fig. 26), the results became somewhat more stable due to the increase in the absolute number of deaths analyzed. The "zeros" plotted on Figs. 25 and 26 may be due to the small populations of these cells.

I performed some multiple regressions of these data to compare the results when the actual socioeconomic data are used instead of "economic levels", and to compare the TSP and SO_x effects when estimated in different ways. In the original paper [90], the census tracts were redistributed to form three groups differing in level of sulfation rate during the heating season. This procedure increased the contrast between groups; however, outdoor exposure to air pollution is generally

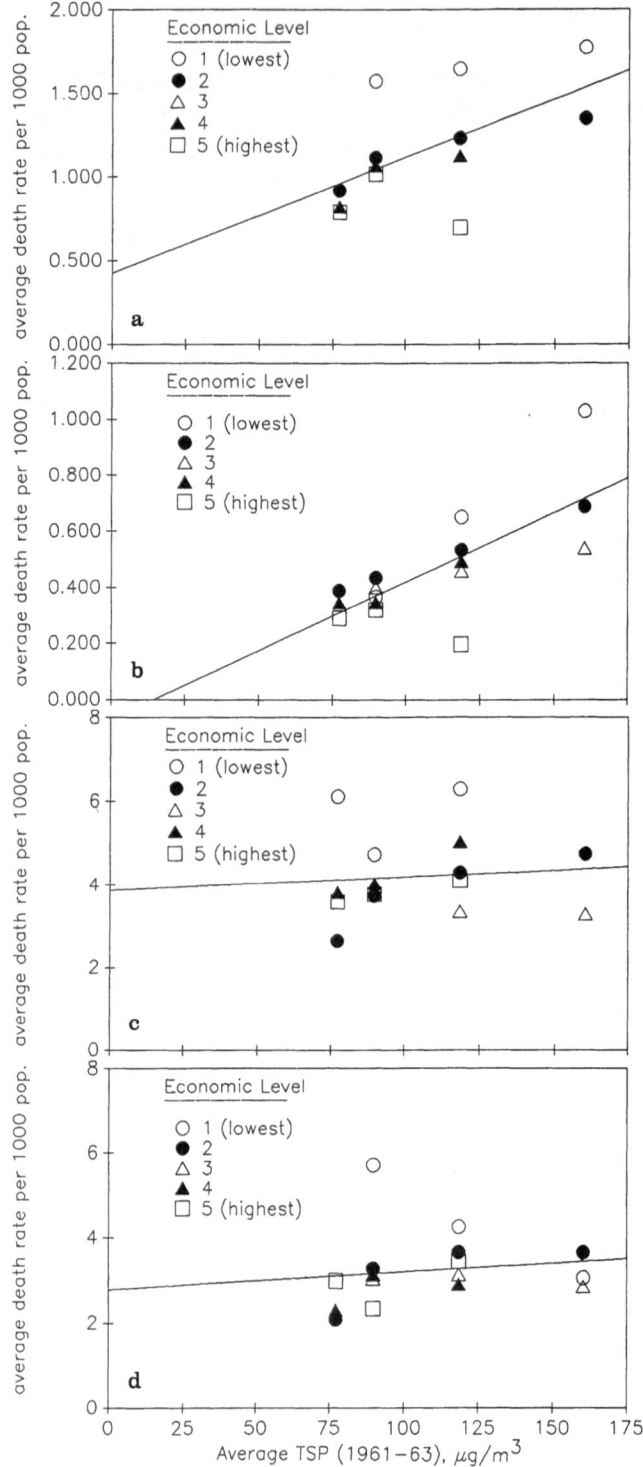

Fig. 22a–d. Arteriosclerotic heart disease mortality in Erie County, NY, 1959–61, by average TSP level. **a** Males ages 50–69. **b** Females ages 50–69. **c** Males ages 70 and over. **d** Females ages 70 and over. Data from Winkelstein and Gay (1970) [90]

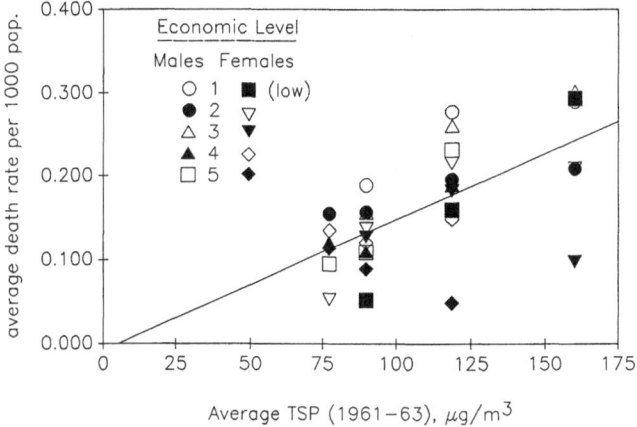

Fig. 23. Cerebrovascular mortality for ages 50–69 in Erie County, NY, 1959–61, by average TSP level. Data from Winkelstein and Gay (1970) [90]

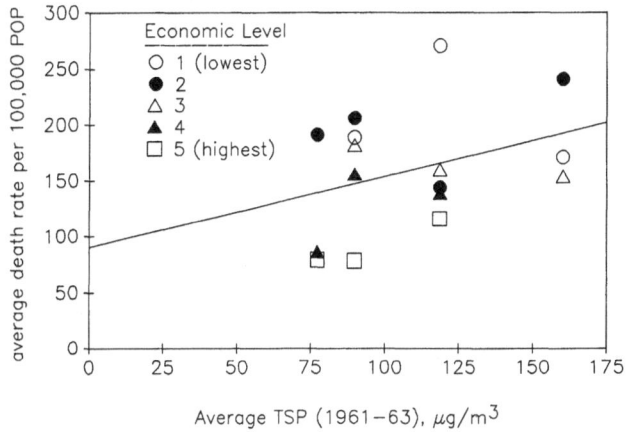

Fig. 24. Respiratory cancer mortality for white males in Erie County, age 50–69, by average TSP level. Data from Winkelstein et al. (1967) [89]

lower during the heating season, so that this grouping may sacrifice realism in exposure. For this reanalysis, I calculated the annual sulfation rates coincident with the four TSP groups [89], and used the annual sulfation rates as the independent variable for the three sulfation-rate groups defined by the authors. The average TSP and SO_x (converted to SO_2) are about 110 and 102 $\mu g/m^3$, respectively; maximum values are 206 and 385 $\mu g/m^3$, respectively and occurred in the same census tract. In addition, Finch and Morris [94] noted that there was potential confounding due to differences among the census tracts in age distribution within the broad age group 50–69, and they presented standardized mortality ratios based on indirect age adjustment, based on the age distributions of each census tract. The correlation matrix for these variables (n = 16) is given in Table 4.

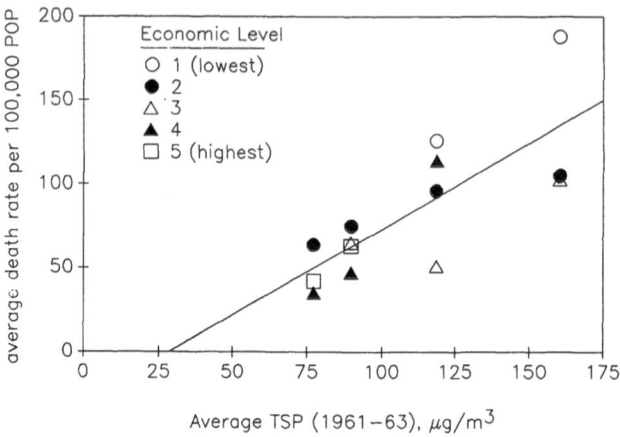

Fig. 25. Chronic respiratory disease mortality for white males in Erie County, age 50–69, by average TSP level. Data from Winkelstein et al. (1967) [89]

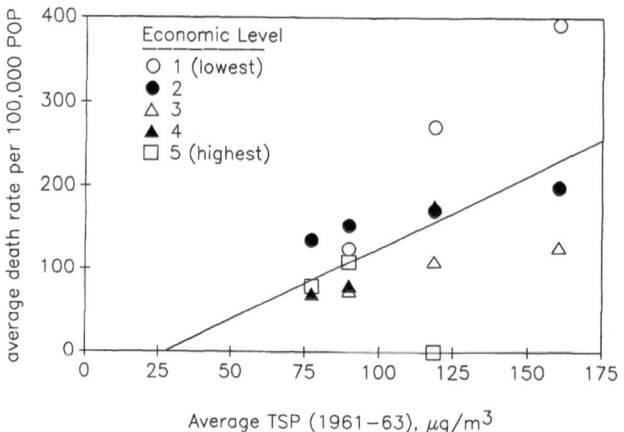

Fig. 26. Mortality for white males in Erie County, age 50–69, with any mention of respiratory disease, by average TSP level. Data from Winkelstein et al. (1967) [89]

Table 4. Correlation Matrix for 1959–61 Erie County, NY Data

SO_x	TSP	Income	Education	% Laborers	Mortality rates (50–69) Male	Female	Av'g
SO_x	0.98	− 0.39	− 0.52	0.69	0.71	0.77	0.76
TSP	1.0	− 0.39	− 0.54	0.67	0.72	0.74	0.75
Income		1.0	0.91	− 0.88	− 0.86	− 0.67	− 0.78
Education			1.0	− 0.79	− 0.80	− 0.68	− 0.76
% Laborers				1.0	0.96	0.84	0.92
Male mortality					1.0	0.90	0.97
Female mortality						1.0	0.98

These correlation coefficients indicate a great deal of collinearity in the data set and suggest that a multiple regression model should use only one pollution and one socioeconomic variable. The finding of similar correlations for male and female mortality suggests the influence of environmental as opposed to personal or genetic factors.

Multiple regressions using various combinations of these variables illustrate some of the pitfalls with collinear data. If "% laborers" is selected as the socioeconomic variable, R^2 is maximized and the pollutant variables lose significance (for TSP, $p = 0.2$; for SO_x, $p = 0.07$). The regression coefficients are of the order of 0.15–0.4% excess deaths/$\mu g/m^3$. If "income" or "education" is selected, R^2 is slightly reduced and the pollutants become highly significant ($p < 0.001$). We can only conclude that is highly likely that either (or both) TSP or SO_x is significantly associated with both male and female mortality. Thus, although sulfation effects on mortality were ruled out by the original analysis using three sulfation levels [89b], sulfur oxides may have contributed to the mortality effects shown in the first paper by Winkelstein et al. [89a]. Considering that sulfation plates do not always provide reliable measures of air concentrations of SO_2 makes it more probable that sulfur oxides could play an important role in explaining the Erie County mortality gradients. Changing the sulfation data from heating season to annual average had no effect on the conclusions of the second paper [89b].

Table 5 compares the TSP regression coefficients. There are five different ways to estimate the change in mortality with respect to TSP:

1. Calculating the slope of the standardized mortality plots in the original paper, using the actual average TSP values for each air pollution level.
2. Calculating the slope from the marginal totals in the two-day contingency tables.
3. Averaging the slopes within the contingency tables.
4. Regressing mortality on TSP (bivariate).
5. Making multiple regressions of mortality on TSP and various socioeconomic variables.

Table 5 shows that there is generally good agreement between the standardized mortality results and the multiple regressions. Most of the other estimation methods overestimate the "true" TSP effect, as might be expected.

The hypothesis of interaction between TSP and economic levels for white males, age 50–69, was tested formally by adding the variable (TSP × economic level) to the multiple regression model. The interaction variable was significant while the variables for TSP and economic level (main effects) lost significance.

The reanalysis of the stomach cancer findings (ages 50–69) showed that TSP was significantly ($p < 0.05$) associated with stomach cancer mortality for both males and females in multiple regressions, and that the socioeconomic variables (income, education, % laborers) were not. The air pollution effect appeared quite strong, with an elasticity of about 1.7 for both sexes. The correlation between sexes was only moderate ($r = 0.51$, $p = 0.05$), suggesting that factors common to a household (such as diet) may have not been overwhelming.

Table 5. Comparison of TSP Regression Coefficients for White Males (deaths/pop./μg/m^3). Data from Winkelstein et al. [89]

Cause of death	Std. Mort Ratio	Cont. Table	Av'g by Econ. Lvl.	Bivar. Regr.	Mult. Regr.
All (50–69)[a]	0.10	0.24	0.14	0.22	0.02[c]–0.07
All (70 +)[a]	0.18	0.36	0.25	0.27	0.18[c]
All (50 +)[a]	—	—	0.15	0.24	0.17–0.19
Resp. Cancer[b]	0.05[c]	0.98	0.32	0.64	0.15[c]
CRD (50–69)[b]	0.94	1.01	1.25	1.0	0.90
Asthma, etc.[b] (any mention)	0.55	1.91	1.13	1.73	1.0

[a] Death rate = deaths/1000 pop.

[b] Death rate = deaths/100,000 pop.

[c] Not statistically significant (p < 0.05)

In follow-up studies in Erie County using ca. 1970 data, Fleissner et al. [95] concluded that the mortality patterns for white males were largely unchanged even though TSP levels had declined about 30%. However, her data showed that all-cause mortality in this age group had declined about 13%, which is consistent with this magnitude of air pollution improvement. Finch et al. [96] looked at changes in both male and female mortality rates and concluded that the longitudinal changes were consistent with the previous cross-sectional gradients. In considering the time-history of air pollution in Buffalo and its effect on mortality, it should be noted that TSP levels in the mid-1950s were about a factor of two higher than during the 1959–61 period studied by Winkelstein et al. [108]. Longitudinal studies should thus consider a longer period of exposure than just the past few years, if truly chronic effects are to be assessed. However, since the middle-age group (50–69) seemed to be most sensitive to the effects of air pollution, there is no suggestion of a cumulative dose effect. Deaths in this age group may reflect the presence of a susceptible subgroup, whose numbers have diminished by the time the cohort reaches age 70.

The Nashville Studies

Zeidberg and colleagues conducted an epidemiological study in Nashville, TN; examining data from 1949 to 1960 on mortality and respiratory disease. Mortality studies comprised respiratory diseases [97], cardiovascular diseases [98], cancer [99] and infant/fetal mortality [100]. This study employed 123 air monitoring stations across the SMSA for sulfation rate and dustfall, and 36 stations for SO_2 (gas) and soiling (COH), for one year (1958–9). Mortality data by census tract were aggregated for the years 1949–60. Socioeconomic factors included median tract family income, education, and the average number of persons per household room. These data were combined into three socioeconomic classifications when analyzing deaths by pollution level. Similarly, three classifications of air pollution were used to examine deaths by economic class. The highest sulfur oxide level was about 35 μg/m^3; the highest soiling, 1.1 COH (about 110 μg/m^3 equivalent TSP), and the highest dustfall level corresponded to about $80 \mu g/m^3$ TSP. Thus, this study characterized exposure well for the year of study, compared to other cross-sectional studies, but it may have been limited by limitations in the ranges of variability in air pollution. Also, smoking habits were not considered, which probably limits the useful results to females.

For respiratory deaths [97], total respiratory deaths were higher for both males and females in the middle economic class for the highest air pollution group regardless of which pollutant defined the group). However, deaths due to lung cancer and bronchitis/emphysema showed the opposite trend, suggesting that smoking may have been negatively correlated with air pollution. If so, the true effect of air pollution on total respiratory deaths may have been even larger. The mortality differences by economic class were large (ca. a factor of 2). Mortality from cardiovascular disease [98] showed more modest differences by economic class (a factor of 1.3) but inconsistent patterns according to air pollution (for the middle class). Total cardiovascular deaths were highest for the highest pollution level for all four pollutants, but only soiling (small particles) showed a linear trend. Deaths from arteriosclerotic heart disease (the largest subgroup, about half the total), showed no air pollution trends. Hypertensive and other myocardial degenerative deaths also showed strong air pollution trends. Females showed mortality trends consistent with air pollution trends (soiling) for all causes of death, suggesting that the trends for males may have been confounded by the implied opposing trends for smoking mentioned above.

In contrast, cancer mortality (all sites) was more uniformly distributed by economic class [99]. Rates for the middle class were slightly lower than for the other two classes. This was also true when rates for the middle class were grouped by air pollution level. However, stomach cancer was significantly elevated by dustfall level (factor of 6). By sex and soiling level, often only the highest level of pollution had a significantly higher mortality rate.

The analysis of fetal and infant mortality [100] concluded there was an association between postneonatal mortality of white infants and sulfur oxides, but that the effect could not be completely separated from socioeconomic factors.

The Nashville study could have examined interactions between economic status and air pollution, as did the Erie County, NY, studies. However, the authors presented only selected "slices" of the possible two-way contingency tables. The study design illustrates one of the fundamental difficulties with cross-sectional studies; small geographic units (census tracts) are required to characterize exposure to locally distributed air pollutants. However, a long period is then needed to stabilize the mortality rates, such that one year's air monitoring record may be inadequate to estimate the cumulative exposure. Also, there may have been substantial residential relocation during the twelve-year period of record.

Mortality Studies by Lave and Colleagues

Air Pollution and Human Health (1978)

Lave and Seskin's cross-sectional regression analysis [101] concluded that about 9% of annual U.S. metropolitan mortality (ca. 1960) was associated with air pollution. These mortality effects were inferred from multiple linear regression analysis of annual mortality rates in the major Standard Metropolitan Statistical Areas (SMSAs) in relation to annual air quality levels (as measured at city centers) and a few other explanatory variables. This study was the first to attempt to characterize the air pollution exposure of an entire SMSA using (often fragmentary) data from a single monitoring station.

Lave and Seskin's work began with an article in *Science* [102] which couched the findings in the general terms of "air pollution". The later publications [101, 107] abandoned this caution and called for emission controls for sulfur oxides on the basis of their findings with respect to air concentrations of sulfate particles. This extension of research results into the policy arena may have been a factor with respect to the ensuing critical reviews [103, 104, 105, 112]. While the Lave and Seskin work broke new ground for cross-sectional studies and included many innovations in methods, because of its many shortcomings and internal inconsistencies, the results should be used with caution. Many of these cautions also apply to similar cross-sectional studies which followed during the 1970s and early 1980s.

Lave and Seskin's handling of the aerometric data generated much controversy. They entered the yearly mean, maximum and minimum 24-hour average values for TSP and sulfates into their regression models, thus defining six pollutant variables. Their rationale for this practice was given as one of ignorance about the "true" specification. However, these six variables are highly correlated with one another, and some may relate to different types of effects (acute versus chronic). To test one type of effect versus another, regressions should have been performed with each pollutant measure entered separately. Moreover, the "1960" sulfate data are a mixed set, consisting partly of 26 biweekly 24-hr samples per monitoring station and partly of quarterly composite averages [103]. Over a year's time, the minimum and maximum values of a set of 26 readings are quite different than those of a set of four readings, thus introducing a bias for a large number of locations in the data set. Further, the sulfate data were not all from the same year, but were taken from 1957 to 1961. The limitations of the sulfate data in 1960–1 are important, since the bulk of the exploratory regression analysis was done with this data set and the "minimum" sulfate variable was selected as the most important measure of sulfur oxides. It is this measure that was most affected by the confusion between biweekly and quarterly samples. However, this was not the case for the 1969 sulfate data set, which was used in a more limited set of analyses [101]. Methods of measurement were inadequate for all of the other pollutants studied by Lave and Seskin, especially NO_3^- and NO_2. The average TSP and sulfate concentrations for the 1960 data set were about 118 and 10 $\mu g/m^3$, respectively. In 1969, these values were about 95 and 11 $\mu g/m^3$, respectively.

Lave and Seskin studied 1960 and 1961 mortality rates in 117 SMSAs, 1969 mortality in 112 SMSAs, and mortality rates in 81 SMSAs common to both data sets. Their dependent variables were 1960 and 1961 total mortality rates (all ages and causes); 1960 mortality rates for four age groups; 1960 and 1961 mortality rates for 15 causes of death; and 1969 total mortality (crude and age-sex-race-adjusted rates). In addition to the six air pollution variables, their independent variables

included population density, population (log of counts), the percentage of population aged 65 and over, the percentage of population classified as nonwhite and the
percentage of population classified as poor. Variables for population migration,
Census region, home- and water-heating fuels, occupations and climate variables
were used at various times. A limited analysis was included of suspended nitrates,
NO_2 and SO_2 (mean, minimum and maximum yearly values in all cases).

Since each metropolitan area was characterized by a single air monitoring
station, the Lave/Seskin study should be regarded as examining the regional
structure of the air pollution/health effects question, as opposed to the local
structure. Of the two pollutants used in the study, sulfate would be characterized as
a regional pollutant, since the typically small sulfate particles tend to travel far from
their original sources. On the other hand, TSP is usually a local pollutant, and high
ambient concentrations tend to be localized around the monitoring site. Thus, if
TSP has a significant association with mortality across an entire SMSA based on a
single monitoring site, it is more likely that the *source* of the particles (such as
heavy industry) is the causal factor through occupational and socioeconomic
factors (which could be diffused throughout the area) than exposure to the heavy
particles *per se* (which will be localized).

Lave and Seskin found both sulfates and TSP to be statistically significantly
associated with total SMSA mortality in 1960 and in 1969. When identical
locations were compared across the two years, there were no significant differences,
although the TSP effect tended to be stronger in 1969 (even though the overall
mean TSP declined from 118 to 95 $\mu g/m^3$). The TSP coefficient values in % excess
deaths per mg/m^3 were 39 and 63, respectively, which are substantially higher than
time-series results but lower than Winkelstein et al.'s results for Erie County, NY
(Fig. 23). If taken at face value, these results suggest that chronic exposure has an
effect well beyond the annual sum of short-term effects (as suggested, for example,
by Evans et al. [106]), and that the poor characterization of exposure in SMSAs
has weakened the ability to detect the effects (compared to groups of census
tracts).

Sulfur oxides tend to have the most definitive pattern (Fig. 27), being widely
distributed throughout the Northcentral and Northeastern regions of the United
States. When Lave and Seskin tested for regional influences by adding a dummy
variable for each of the nine Census Divisions (regression 3.5–9 in Ref. [101]), the
sulfate coefficient (minimum values) dropped 40% and lost statistical significance;
the TSP coefficient (mean values) increased by 24%. Large decreases in the sulfate
coefficient also occurred when population migration was taken into account. The
authors described these effects as ". . . pollution coefficients . . . not substantially
affected . . . ."

When nonlinear pollution specifications were examined (Figs. 3.2, 3.3, 7.1, and
7.2 in Ref. [101]), the authors discounted the shapes of the resulting dose-response
curves (which suggested a threshold for TSP at 110 $\mu g/m^3$ in 1960 and at
3.5–5.5 $\mu g/m^3$ for minimum sulfate in 1969), emphasizing instead statistical tests on
the superiority of the overall fit of the model. This approach established the linear,
no-threshold model as the specification of choice unless it could be shown that a
more complicated model was statistically superior. The use of a threshold is an

Fig. 27. Suspended sulfate aerosol patterns in the U.S., based on ASTRAP model output [112]. Source: Lipfert et al. (1988) [3]

important distinction with respect to air quality standards and requires more sophisticated model comparison tests (see below).

By cause of death, the most significant air pollution effects were for cardiovascular disease (both sulfate and TSP) and total cancers (sulfate). Stomach cancer was not associated with TSP, and respiratory cancer was not associated with either pollutant. Deaths from respiratory disease were not associated with sulfate, but were weakly associated with minimum TSP. This latter association was stronger for tuberculosis and asthma; influenza, pneumonia and bronchitis were not associated with either pollutant. Lave and Seskin also examined some variables not plausibly associated with air pollution (suicide, venereal disease, and crime rates) to check for spurious (non-causal) associations; none were found. However, these non-causal findings were assigned a non-causal interpretation, whereas findings of nonsignificant air pollution effects with similar R^2 values for respiratory diseases were ascribed to "sampling variability". Sampling variability could have been a factor with suicide or VD as well.

By age, sex, and race, air pollution effects were stronger for nonwhites and for females, for whites from age 45–64, and for nonwhites 65 and over. For most of these subgroups, TSP was more important than sulfates. TSP was also important for nonwhites younger than 45, for which nondisease deaths tend to dominate. Since age-race and disease groups were investigated only separately, additional caution is advised when considering causality.

The authors concluded from these regressions: ". . . levels of certain pollutants in the air (as they prevailed in certain cities during the period 1960–9) *caused* increases in mortality, and presumably in morbidity as well . . ." (my emphasis). However, the design of these studies and the data they used make even the published statistical associations suspect.

There are several flaws in the Lave/Seskin studies in addition to their problems with the air quality data:

1. The regression model used was derived on an *ad hoc* basis, without a well-specified *a priori* hypothesis, and without optimization or consideration of alternative socioeconomic variables.
2. Important demographic variables and personal risk factors were omitted from the analysis, including smoking habits, diet, and education. Subsequent analysis by others (see below) showed all of these factors to be important and to be correlated with pollution variables.
3. The authors' conclusions are inconsistent with their own analysis. In nearly every case, adding explanatory variables to the data set decreased the significance of the pollution variables, especially for sulfate. Pollution was not significant for respiratory causes of death, in contrast to most of the previous studies reviewed above. The authors ignored these indications of non-causal associations.
4. The age groups used to analyze age-specific mortality were too broad (45–64, 65 and over). Age adjustment is still needed within such groups.
5. For some pollutants, the SMSA is too large a political subdivision for estimating representative air pollution exposures from monitoring data. SMSAs are usually groups of whole counties and may have only one monitoring station, usually in the central business district. Thus, no information is provided on exposures in the suburbs, where often the bulk of the SMSA population lives. Where multiple stations were available for an SMSA, they were apparently not used in this study.
6. The study included a cross-lagged analysis of all combinations of 1960 and 1969 mortality, air pollution, and socioeconomic variables, and found that 1969 mortality was better predicted by 1960 air pollution than by data from the same year. However, since the 1960 sulfate data were so badly flawed (as used in this study), this finding is meaningless.

The Lave/Seskin studies were among the first to apply econometric methods to the health effects of air pollution. As a result, their emphasis was on regression analysis, which imposes a linearized structure on the data and makes the analysis of interactions more difficult. They made no attempts to explain their results from a physiological perspective, yet they called for revisions to the health-based standards as a result of their findings. In their summary, they stated that their study was focused on "... how human mortality is affected by a *reduction* (my emphasis) in air pollution". However, a cross-sectional study only examines variations by place. To apply the spatial differentials and regression coefficients to change over time requires the establishment of a causal model, including physiological explanations. Such a model should be verified by application to actual case histories of pollution abatement.

The 1982 Study (Chappie and Lave [107])

Partly as a result of the extensive criticisms of the earlier work and the publication of contrary findings [108, 109], Lave and his colleagues performed a similar cross-sectional analysis using more recent data and attempted to address some of the criticisms. The 1974 mortality rates for about 100 SMSAs were analyzed, and the regression coefficients were compared to Lave and Seskin's results for 1960 and 1969.

The dependent variables were 1974 total mortality and mortality from disease (total less external causes) for SMSAs. There was also a limited analysis of corresponding data for counties and cities. The independent variables followed the previous study (population density, population [log of counts], % > 65, % nonwhite, % poor, sulfate particles [mean, minimum and maximum], TSP [mean, minimum and maximum]); income, % college graduates, tobacco and alcohol expenditure per capita, occupation, nutrition variables, physicians per capita and birth rate were used in addition. The 1974 mean air pollution values for sulfate and TSP were 9.6 and 75 $\mu g/m^3$, respectively.

Chappie and Lave found that in 1974 the sulfate effect became much stronger (accounting for about 12% of total mortality) and the TSP effect dropped to almost zero. Eliminating non-disease causes from the dependent variable slightly strengthened the sulfate effect; adding an education variable weakened it and adding variables for smoking, alcohol use and occupation had little effect. When regressions for SMSAs, counties, and cities were compared (96 locations), there were slight reductions in the sulfate effects and small nonsignificant increases in the TSP effects, but the smoking variable became nonsignificant and negative in some cases. The authors concluded: "A strong, consistent and statistically significant association between sulfates and mortality persists . . . . These results can be used to support stringent abatement of sulfur-oxides air pollution . . . ."

It is unfortunate that the 1982 study was structured as a defense of the earlier Lave and Seskin work rather than as an independent replication. It might have been more interesting to derive an independent regression model from the later data, and then to compare the air pollution effects. Since geographic patterns in U.S. mortality have been stable over the years [109], one would expect consistent regression results over time. However, most air pollution levels have decreased since the 1960s, and it would be interesting to determine to what extent these changes are reflected in the regression results. It is particularly important that the use of mean, minimum, and maximum measures of each pollutant concentration were retained, since this algorithm was apparently derived from the flawed 1960 sulfate data originally used by Lave and Seskin [101]. No statistical or physiological justification has ever been advanced for this procedure, and most authors who have reanalyzed these data elected to use only the mean pollution values.

Chappie and Lave showed that the 1974 regression coefficients were not statistically different from the 1960 and 1969 coefficients; however, their model allowed each year to have its own intercept, which suggests that important factors are missing from the model. Year-to-year consistency in the data was shown by this analysis, which was expected. In most cases, adding new variables to the specification and using only one measure rather than three for each pollutant reduced the importance of the sulfate variable. Eliminating external (non-disease) causes of death had the opposite effect, since it is the spatial pattern of heart-disease mortality that appears to be associated with sulfate concentration patterns.

The tobacco and alcohol consumption variables ($/capita) used by Chappie and Lave are poor surrogates for actual consumption, since they do not account for price variations, for sales to nonresidents (due to differences in state/local taxes, for example), and are based on self-reported data on retail sales which suffer from varying degrees of underreporting. This criticism is borne out by the weak effects shown for these variables in this study, in contrast to previous findings [108]. Thus, including these quasi-surrogate variables does not constitute "controlling" for smoking or drinking, as the authors claimed.

The effects of smaller aggregation units (counties and cities versus SMSAs) were also examined with all six pollution variables and with total mortality rates. This procedure should have been extended to the cases with only two pollution variables and with non-external causes of death (external causes are likely to be higher in industrial [i.e., polluted] cities, but the effect would be lost in an SMSA analysis when suburbs are included). R^2 values were not as high as some other studies [108], and some variables had coefficients of the "wrong" sign, all of which indicate problems with the specification (i.e., model) or perhaps input data. Part of the advantage of using cities rather than SMSAs is to increase the number of observations and to evaluate robustness. Chappie and Lave examined only one set of cities, the central cities corresponding to the SMSAs. However, many SMSAs have more than one city; in the 1980 Census, there were about 300 SMSAs and over 900 cities (with greater than 25,000 population).

The results of Chappie and Lave are presented for *a priori* specified models. A stepwise exploration procedure would give insight into the covariance structure and the extent of the collinearity problems. Birthrate was only used as a possible surrogate variable explaining the demand for medical care. However, a better use of birthrate data is as a predictor for mortality to represent possible undercounting of the population, which can inflate both mortality and birth rates (by the same percentage). This is especially important when a mid-census data year is used (1974), since population errors may be considerable especially in poorer (and often more polluted) areas. The diet (nutrition) data used were for only four regions of the country and were out of date (1955). Finally, Chappie and Lave presented no analysis of residual patterns or of possible spatial autocorrelation.

Comparing the Lave/Seskin and Chappie/Lave pollution coefficients across years, the TSP coefficients dropped significantly, which may be consistent with the substantial drop in the ambient TSP concentrations over this period. However, the sulfate coefficients *increased* substantially (more than doubled), in spite of a *drop* in average ambient air concentration. The net total pollution effect on mortality appears to have remained approximately constant over the years, and the proportions for the two pollutants have shifted. Since no reasons for such a change have been advanced based on statistics, air quality measurements, or physiology, a possible conclusion is that the indicated pollution effects on mortality are in fact surrogates for other phenomena.

The Chappie/Lave analysis sheds no light on whether the effects claimed are long-term (chronic) or just the sum of acute exposures over a year. This might have been resolved with a cross-validation analysis, which could provide insight into the physiological situation. Although the Chappie/Lave study is more complete than the Lave/Seskin work in some respects, no other air pollutants were explored (such as CO or ozone, both of which might be expected to be more physiologically important for heart disease than sulfates); drinking water hardness was not considered, nor was population migration.

Cross-Sectional Studies by Lipfert and Colleagues

Preliminary Studies

Total mortality from 1969 was analyzed in up to 136 U.S. cities [110]; in addition, the findings for 60 SMSAs were compared to their central city counterparts. The independent variables in these regressions included: % > 65, % nonwhite, % poverty, population density, birth rate, % housing built before 1950, mean and maximum SO_2, mean and minimum SO_4^{2-}, TSP, iron, manganese and benzo(a)pyrene. Stepwise regression optimizations were used to define the "best" models. The main objective of this study was to point out some of the pitfalls in this type of analysis; I reached the following conclusions:

1. Statistical significance of SO_4^{2-} depends on which other variables are included in the model.
2. The significance of the pollutant variables was sensitive to the use of cities versus SMSAs as the observational unit.

This preliminary study did not include a variable for cigarette smoking and used only one year's mortality data, which could be unstable in the case of smaller cities (see Refs. [80] and [108]). The use of housing age as a surrogate for housing adequacy was questionable, and may serve mainly to illustrate the collinearity between sulfate and other socioeconomic indicator variables for the Northeastern United States. The instability of the socioeconomic coefficients between cities and SMSAs suggested that important variables were omitted in both specifications.

1959–71 City Mortality Studies

The main body of analysis of city mortality for 1959–71 has been published in several formats [4, 103, 108, 110]. It consists of analysis of 1969–71 mortality in up to 201 U.S. cities and a limited analysis of 1959 mortality in a subset of these cities. Total mortality was analyzed for both periods; for 1969–71, mortality by 10-year age groups and by broad cause of death groups was also examined.

The independent variables included % > 65, % nonwhite, % poverty, birth rate, % old housing, smoking, education, population density, population change, region of residence, mean concentrations of SO_4^{2-}, TSP, iron, manganese, and benzo(a)pyrene [B(a)P]. The 1969–71 mean values of sulfate and TSP were 9.1 and 86 $\mu g/m^3$, respectively; in 1959, they were 11.1 and 113 $\mu g/m^3$; the average TSP for the years 1953–57 was 149 $\mu g/m^3$.

The overall conclusion was that there was a small ($\sim 5\%$) but statistically significant association between air pollution and urban mortality which was stronger in certain cases:

1. in 1969, compared to 1959
2. in the Northeast and North Central States

3. for persons aged 65 and over
4. for non-specific causes of death (heart disease)
5. for manganese and TSP air pollution.

Sulfate was significant (positive) only for older age groups (75 +) and only when regressed as the sole pollutant with an *a priori* model specification (as opposed to stepwise regression). In general, the analysis of the health effects of sulfate may have been compromised by multicollinearity with both other pollutants and socio-economic variables. No thresholds were suggested for SO_4^{2-} or Mn; a threshold of 85–130 $\mu g/m^3$ was suggested for TSP (1969–71). Regressions with lagged air pollution data were also compromised by being based on fewer observations, but they suggested that the lagged variables were often slightly superior. Regressions comparing geographic units (cities, counties, states) showed that as the size of the unit increased, TSP lost significance and sulfate gained significance (especially at the state level).

These studies were limited by the failure to study adequately long-term exposures and by omitting other potentially confounding variables such as drinking water hardness and diet. Other criticisms include:

1. The use of the age of housing as a surrogate variable for the adequacy of housing was questionable; however, this variable demonstrated the sensitivity of the sulfate variable to model specification, especially for total mortality.
2. Drinking water hardness, ozone, and migration should have been included as independent variables.
3. The spatial autocorrelation of regression residuals should have been examined.
4. The results for manganese may have been overinterpreted, and most likely are a surrogate for occupational exposure. Analysis by sex would be helpful to verify this hypothesis. A physiological hypothesis is lacking for the effects of this pollutant at low concentrations.
5. Given the availability of air quality data for some species back to 1953, an attempt should have been made to construct a lifetime exposure index to examine very long-term effects. Analysis of mortality differentials in a given year cannot distinguish between chronic effects and the sum of acute effects [106].

This analysis showed that for age-specific mortality, in most cases *neither* SO_4^{2-} nor TSP was significant. If the manganese effect were a surrogate for occupational effects, then there would be essentially *no* effects of pollution on mortality, a point that was overlooked by many who have cited this work. By implication, the need for better age adjustment is implied when total mortality is analyzed.

1969–70 SMSA Mortality

The next major publication was a reanalysis of Lave and Seskin's 1969 total mortality data set for 112 SMSAs, using corrected data and many new independent variables [80]. The purpose was to provide a more rigorous analysis of the portion of the Lave/Seskin data that had usable environmental data, and to compared it with the original analysis by Lave and Seskin [101].

The dependent variables were 1969 total SMSA mortality from Lave and Seskin; total mortality, mortality by sex and for two broad age groups, all for 1970 (to correspond with the Census). The independent variables were: from Lave and Seskin: % > 65, % nonwhite, % poor, population density, log of population, mean TSP, mean and minimum SO_4^{2-}. Additional new variables included: % Negro, % other nonwhite, % net migration, diet, smoking, % coal heat × degree-days, % wood heat × degree-days, drinking water variables, ozone, iron, and manganese.

The main conclusion that I drew from this analysis of 1970 SMSA mortality was that pollution effects on mortality were sensitive to both model specification and the data set used. Specifically:

1. Minimum SO_4^{2-} had no advantage over mean SO_4^{2-} and was less significant, other things being equal. When a complete model specification was used, sulfate was rarely significant (as either measure). These cases occurred more often for females. TSP was more important for deaths to those under 65 years of age.
2. Drinking water quality, ozone, migration, and racial variables all had important effects on the regressions. Only when I used a complete model specification did the coefficients for age, race, poverty, and smoking assume values consistent with exogenous estimates; this result implies that these additional variables were necessary.
3. Since manganese was only important for males (65 and over), this variable may be a surrogate for long-term occupational effects. This suggests that the city mortality studies should be replicated by sex of decedent.
4. The analysis was incapable of distinguishing between linear and threshold models.

Because data for all the new variables were not available for all 112 SMSAs, this study had problems with missing data, depending on the model specification. Thus, I could not determine whether sulfate lost significance because of the addition of new variables to the data set or because of the deletion of the observations not having data on the new variables. The sensitivity of the results to age grouping implies that finer age groups should be explored.

Regardless of the missing data problems, the study showed that sulfate effects are not robust and suggested that the manganese effect was probably a surrogate effect rather than a bona fide effect of air pollution on mortality.

1980 City Mortality

In the most recent work [3], we performed a statistical analysis of spatial patterns of 1980 U.S. urban total mortality (all causes), evaluating demographic, socio-economic and air pollution factors as predictors. Specific mortality predictors included cigarette smoking, hardness of drinking water, heating fuel use, and annual concentrations of the following air pollutants for 1978–82: ozone, carbon monoxide, sulfate aerosol, particulate concentrations of lead, iron, cadmium, manganese, and vanadium, as well as total and fine particle mass concentrations from the inhalable particulate network (dichotomous samplers accepting particles less than about 15 μm diameter with a further size cut at about 2.5 μm). In addition, based on output from the ASTRAP long-range transport dispersion model [112], estimates of sulfur dioxide, oxides of nitrogen, and sulfate aerosol concentrations were made for each city and entered into the analysis as independent variables. Because the number of cities with valid data on air quality and water hardness varied considerably by pollutant, we considered several different data sets ranging from 48 to 952 cities. The mean sulfate concentration was about 5 μg/m^3; the mean total (inhalable) particle mass was about 40 μg/m^3. Note that this measure of total particle mass is probably more comparable to British smoke than TSP. These mean values represent substantial reductions from those of previous studies.

The regression models and data were tested for outliers, influential observations, and behavior of residuals. The models which passed these tests showed sulfate aerosol, iron particles, and (to a lesser extent) total particle mass to be associated with mortality, although, depending on the data set considered and the choice of variables or factors to account for demographic effects on mortality, there were variations in the specific pollutants which showed statistical significance.

These results were compared to previous results and data for 1970. We found that 1970 pollution was not a better predictor for 1980 mortality than 1980 pollution. The only coefficients which underwent a statistically significant change since 1970 were those for poverty and the percentage of population 65 years of age or older (borderline significance for the latter). Since sulfate was not significant in 1970, the large standard error found for that data set does not preclude the 1980 coefficient for that period.

Since the computed sulfate variable was so highly significant in the developmental regression models, the analysis was repeated using computed NO_x and SO_x data as alternative independent variables. These variables were also highly significant predictors for all-cause mortality. Because of the high degree of similarity of alternative regression models employing different pollutants as predictors, three different bases were employed to compare regression models. Using these varying criteria, SO_2, sulfate aerosol, and manganese particles all appeared to provide better regression fits than the other pollutants.

A two-stage procedure was used employing mortality "adjusted" for the non-pollution variables (demographics, smoking, poverty). The adjusted mortality rates were computed as the residuals from a first stage regression including all the non-pollutant variables. This procedure was carried out twice, with and without drinking water hardness in the model. These residuals were then regressed against each pollutant variable in turn as a bivariate procedure, using only the cities for which the pollutant data were available. This protocol assured that all the different data sets defined by the availability of pollutant data were treated alike with respect to non-pollutant variables, but carried the risk that some portion of the pollutant effects were inadvertently included in the "adjustment" procedure due to collinearity between demographic and pollutant variables. The differences in fit among the various sulfate measures, SO_2, NO_x, Fe, Mn, fine particles and total particles were minimal, as seen in Fig. 28. The correlation coefficients and their (positive) 95% confidence limits are shown in Fig. 28a; the metal particles (iron and manganese) showed the highest values consistently for both data sets. The elasticities[3] are

---

[3] Elasticity is a nondimensional regression coefficient defined by

$$e_i = b_i(\bar{x}_i/\bar{y})$$

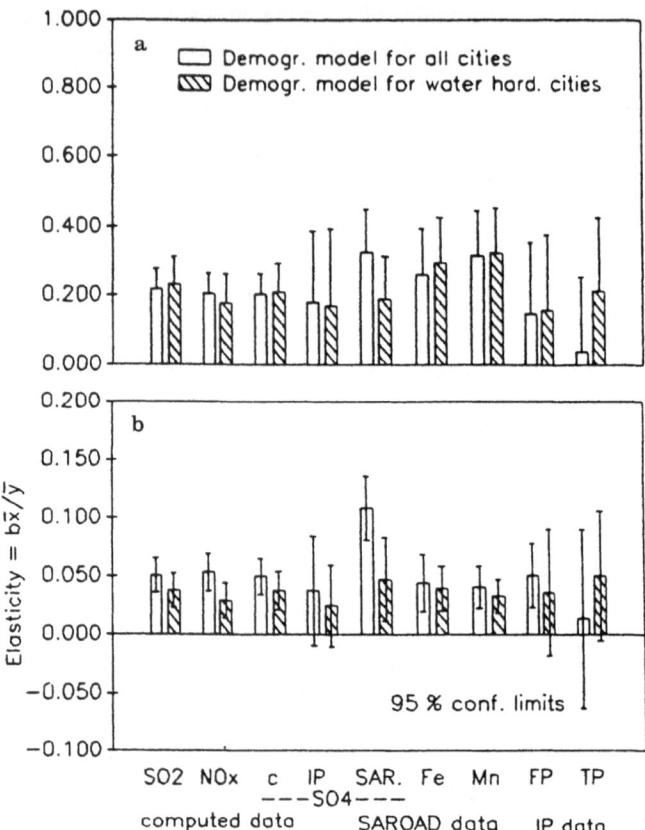

Fig. 28a, b. Correlation coefficients **a** and elasticities **b** for a two-stage regression analysis of 1980 total mortality vs. various air pollutants in U.S. cities. Source: Lipfert et al. (1988) [3]

compared in Fig. 28b; most of the values are between 3% and 5%, with the exception of measured sulfates for the all-city data set (without water hardness).

Our inability to identify a single pollutant most closely associated with mortality rates may be interpreted in two ways. Since this was a total mortality analysis, it is possible that all the pollutants are associated in some way with either a component of the population or some particular cause of death. It is also possible that none of them is causally related to mortality and that they all reflect the effects of some additional, unmeasured variable. Thus, we concluded that statistical criteria alone are not sufficient to define air pollution-mortality relationships.

Other U.S. Cross-Sectional Mortality Studies

Ridge Regression Methods

McDonald and Schwing [113] and Schwing and McDonald [114] applied ridge regression methods to this problem, as a way to deal with the multicollinearity (non-orthogonality) typically found among the independent variables required for a complete mortality model. Ridge regression is a technique which relaxes the least-squares constraint by introducing an arbitrary parameter, k, to the diagonal entries of the correlation matrix, which biases the estimators. When k = 0, the least squares solution obtains. Typically, multiple runs are made for varying values of k and the regression coefficients are plotted versus k. A judgment must then be made as to the value of k for which regression estimates have "stabilized". With severe collinearity, the coefficients may change sign as k is increased from zero. Confidence limits are not possible for these biased coefficients.

McDonald and Schwing's studies had several other interesting features; in the first paper, air pollution "potential" was used in lieu of concentration measurements. The potential was based on emission density, SMSA dimensions and dispersion characteristics, and thus freed the data from biases due to locations of monitors and from measurement problems. However, this approach cannot handle the transport of pollutants between SMSAs, which may be important for sulfur compounds (long-range transport models such as ASTRAP account for this phenomenon). The second paper included variables representing cigarette smoking and background ionizing radiation. Both studies used age-adjusted or age-stratified mortality rates.

McDonald and Schwing [114] used a data base of 60 SMSAs and total age-adjusted mortality rates for 1960. They found substantial instability when 15 independent variables were used (which is not surprising given only 60 observations), but the regressions using six variables were more stable. In both cases, SO_2 potential was an important variable, with a standardized coefficient of about 0.25, contributing about 1.5% of the mean mortality. The R^2s for these models were around 0.7, substantially higher than achieved by Lave and Seskin [10] with 1960 age-adjusted data. No measures of particle concentrations were included in this initial study.

Schwing and McDonald [114] expanded the variable list to include estimates of cigarette smoking, some measured air pollutants (1965 or 1969 values) and background ionizing radiation. However, they had data on only 46 SMSAs. They studied 1959–61 mortality rates for specific age groups and causes of death for white males, and presented results for least-squares, ridge, and sign-constrained (minimum residual sum of squares) regressions. Although the sulfate and nitrate contents of TSP were included as independent variables, no general measures of particle loading were included. The results of this study were highly variable for air pollution, in part because several different measures of the same pollutant were often used simultaneously and because the number of observations was small in comparison to the number of independent variables. The authors reported that the total elasticity for all sulfur compounds on all-disease mortality was about 2–3% for ridge regression and 3–4% for sign-constrained least-squares. Corresponding elasticities for smoking were in the range 7–11%. When compared to the corresponding results from Lave and Seskin [101], the regression coefficients (elasticities) of Schwing and McDonald were generally substantially lower.

Thomas [115] also found air pollution elasticities in the range of 2% for 1960 mortality, using ridge regression and measured values of both sulfates and TSP.

I also used ridge regression with 1970 SMSA mortality [80]; the sulfate coefficients were unchanged and in one case, the TSP coefficient did not reach a stable value with increasing k.

On the basis of these various studies, I concluded that the ridge regression technique offered no particular advantage.

Mendelsohn and Orcutt [116]

These authors studied mortality for the entire United States for 1970 by defining 404 groups of contiguous counties. Air quality was based on a limited set of measurements and interpolated estimates, for 1974. Death certificates were merged with the Census Public Use Sample to supply more details on individual versus group demographic characteristics, a major contribution of this study.

The dependent variables were 1970 mortality rates grouped by sex, race, and broad classifications for age and causes of death. Independent variables included (1970) age, income, education, and marital status, by sex, race, and broad age groups (24 data cells); county average age, education, income, number of children, marital status, employment status, housing status, migration, central city residence, climate and regional residence. The pollutants included estimated annual means for SO_4^{2-}, NO_3^-, TSP, CO, SO_2, NO_2, and ozone. The authors concluded that sulfate was ". . . closely associated with many deaths; . . ." and that ozone and NO_2 had no apparent effect on expected lifetimes.

The chief methodological contribution of this study was the matching of data on individuals with group-mean data, a partial response to the "ecologic fallacy" criticism. The study thus compared the effects of certain demographic variables on mortality averaged across all residents of a county with the effects averaged for age-race-sex specific population cells. The results showed that the county-wide variables were still important in the regressions, even when entered along with the variables which pertain better to individuals [117]. This counterintuitive finding may indicate that these individual variables were surrogates for some other effects on mortality rather than just aggregate averages of individual characteristics. If so, the arguments against literal interpretation of the results of "ecologic" analyses would be strengthened, a point that these authors overlooked.

Among the main criticisms of the Mendelsohn and Orcutt study is the use of 1974 air quality data to "explain" 1970 mortality data. Concentrations of most primary pollutants declined substantially from 1970 to 1974 (1974 mean sulfate and TSP were 10.5 and 70 $\mu g/m^3$, respectively) and the preferred air quality measures would be averages for say five-ten years *before* the date of death, not four years *after*. In addition, it is practically impossible to estimate exposure for the entire United States on the basis of existing monitoring networks, and thus this study may be affected by spatial autocorrelation, especially because of the process used to estimate air quality data where measurements were lacking. The one possible exception to these criticisms might be for sulfates, which tend to be distributed more uniformly spatially and which have not changed as much over time as have most primary air pollutants. However, the statistical significance of the sulfate variable could be enhanced relative to other pollutants just by virtue of having less measurement error. In addition, the omission of smoking and drinking water quality (hardness) may have overstated the effects of sulfate, given the regional distributions of these variables and previous findings of their significance by others [80, 108, 111].

This study showed that there was no particular benefit from a more rigorous approach to estimating the effects of socioeconomic variables on mortality, and thus it tends to validate the use of cruder methods, at least on a relative basis. The findings on air pollution effects on mortality must be viewed with caution because of the questionable procedures used to estimate air quality exposures. In addition, many of the R^2 values given are seriously overstated [118].

Selvin et al. [119]

Analyzed total (all causes) mortality by county for 1968–72, for both 3082 individual counties and 410 county groups according to the 1970 Census Public Use Sample (see Ref. [109]). The dependent variables were mortality rates for all causes for white males and white females, ages 45–54. The independent variables included: land area, population, net migration % change, % urban, % black, % foreign stock, divorce rate, % < 5 yrs. old, % > 65 yrs. old, % income < \$3000, % income > \$15 000, % 4 yrs. college, occupational variables, housing variables, climate variables, elevation, median 1974–6 TSP, median 1974–6 SO_2 and median 1974–6 NO_2.

The authors concluded that they found ". . . no persuasive evidence of a link between air quality and general mortality levels . . ." and that causal relationships cannot be inferred from this type of analysis.

The Selvin et al. study is a variant of that by Mendelsohn and Orcutt [116], but retained many of its problems:

1. The air quality data postdate the mortality data and were based on interpolated values.
2. The study erroneously assumed that it is sufficient to analyze only one age group (since the average rates for other groups are exponentially related to one another), and that it is necessary for all the variables to have normal distributions.
3. Residuals were examined for heteroscedasticity but not for spatial autocorrelation.
4. Air quality data were pooled without regard to measurement method, which could give a bias for SO_2 and NO_2, assuming some stations used bubblers and others used continuous electronic instruments.
5. In their discussion, the authors ignored the findings of statistically significant SO_2 coefficients for the whole country and the South. They presented no detailed comparisons of their results with those of Mendelsohn and Orcutt, in spite of the strong parallels between the two studies. The elasticities for SO_x appear to be quite similar for the two studies.
6. Suspended sulfates and ozone should have been included in the list of variables, in part to check previous findings and because their spatial patterns tend to be smooth and thus missing values would be easier to estimate.

Crocker et al. [109]

Crocker and his colleagues at the University of Wyoming studied total mortality and certain diseases in 60 U.S. cities for 1970. Their analysis featured a simultaneous equation approach to estimating the effects of medical care on mortality, a factor which had not been considered by most previous authors. Their dependent variables were 1970 mortality (all ages) for all causes, for pneumonia and influenza, and for early infant diseases. Independent variables included age, income, education, smoking, diet, medical care, climate, and mean concentrations of TSP, NO_2, SO_2. Sulfate was not included in this study.

The authors found no pollutants to be statistically significant for total mortality, TSP significant for pneumonia and influenza, and SO_2 significant for early infant diseases. A cost-benefit analysis was performed with these results.

Unfortunately, there were substantial flaws in the air quality data used which prevented this study from reaching meaningful conclusions (in addition, given the number of independent variables, 60 observations [locations] are insufficient for reliable results). SO_2 was the only measure of sulfur oxides and these data were obtained from bubblers, which tend to give low readings. For example, the SO_2 value used for New York City was too low by a factor of four. The data reported for many cities were below the detection limit for this method. Measurement noise of this type will bias the regression coefficients downward. The NO_2 data used in this study were based on the Jacobs-Hochheiser method, which is no longer considered valid because of substantial interferences (biases). The data base used mixed mean and median (50th percentiles) measures in the same data set, which amounts to biasing part of the data set downward. The pollution data cited show anomalous values for several cities.

The dietary data were from 1955, not from 1965 as reported. Thus they are too outdated to be meaningful as a 1970 health predictor. The simultaneous equations method used to study the effect of medical care was not very successful in predicting mortality ($R^2 = 0.39$) and nearly cancelled the effect of smoking, since this variable appeared in the complete specification twice, with opposite signs.

Gerking and Schulze [120]

Continued the analysis of 1970 total mortality in 60 U.S. cities using ordinary and 2-stage least squares procedures (apparently the same data set used by Crocker et al. [109]), emphasizing the simultaneous equations approach. They concluded that the estimated effects of air pollution are highly sensitive to model specification and, as a result, relatively little is known about the effects of long-term low-level air pollution exposures on human mortality.

While the above conclusions may be true, the points would have been more convincing had a larger data set been used, comprised of better quality data; results are usually conflicting when faulty data are used. (see comments above on Crocker et al. [109]).

Namekata et al. [121]

Namekata and his colleagues at the University of Illinois (at Chicago) performed a two-phase study of air pollution in the City of Chicago from 1971–5. The first phase was cross-sectional and studied differences among 76 neighborhoods (community areas) for several causes of death (age adjusted), averaged over the five years. The second phase examined daily variations averaged over the 76 areas for total non-accidental deaths and heart disease. The pollutants of interest were TSP and SO_2. Socioeconomic variations were accounted for through the use of three index variables which were developed by an outside agency. These variables included income measures, housing status, and education; the racial composition of the population was not specifically accounted for, but the authors reported that it was highly correlated with these three socioeconomic factors and thus was accounted for indirectly. Smoking habits were not accounted for. This study affords an opportunity to compare the results of cross-sectional and time-series regressions.

The cross-sectional regressions found about 25% excess deaths due to TSP, mainly in heart disease. The regression coefficient is a factor of 3-4 larger than was found in cross-sectional studies between SMSAs or cities [101, 108]. No independent effects were found for SO_2; when both pollutants were entered, the SO_2 coefficient was negative. Although the magnitude of the TSP effect seems quite large considering the mean value of 83 $\mu g/m^3$, it is possible that the effects of historical exposures were still being felt (previous TSP levels were 2-3 times this figure). As a rough check on the plausibility of an effect of this magnitude, we note that the total non-accidental age-adjusted mortality for Chicago was about 33% above the national average, taking into account the racial composition of the population. It is therefore possible (though unlikely) that most of this difference is due to air pollution. Heart disease deaths accounted for most of Chicago's excess mortality with respect to the national average.

The indicated pollution effects in this study were probably confounded with either racial differences, with smoking habits, or with both. Examining the results for diseases which may be associated with these factors can give help in interpretation. With regard to confounding by smoking, we note that emphysema deaths were strongly associated with TSP, but that lung cancer was not. This comparison must therefore be regarded as inconclusive. "Marker" causes for blacks may include diabetes (weakly associated with TSP), cerebrovascular disease (not associated with TSP), cirrhosis of the liver (strongly associated with TSP), pneumonia (not associated with TSP) and cancer of the genito-urinary organs (not associated with TSP). This comparison is also equivocal but certainly does not suggest strong confounding.

The daily mortality study used no filtering or seasonal corrections but employed average temperature in the regressions along with dummy variables for the days of the week. Air pollution was not measured every day, so that the analysis was limited to those dates with at least 20 TSP or 18 SO_2 measurement sites. These numbers of sites represent a substantial improvement over the estimates of daily exposure made for most other time-series studies. As discussed above, a model of this sort is likely to predict more excess mortality for air pollution than one which employs filtering or seasonal adjustment, since a single temperature variable cannot account for both heat waves and the normal seasonal cycle in mortality. Either TSP or SO_2 (separate regressions) accounted for about 7% of Chicago daily mortality. When the product of $SO_2 \times$ TSP was entered (alone), the result was less significant and accounted for about 3% of mortality. Lags of three and six days gave lower results, as did regressions on heart disease deaths. Since these regression results are probably high estimates with respect to a more complete time-series model, I concluded that the cross-sectional or chronic mortality effects appear to substantially exceed the daily (short-term) effects in Chicago, the shortcomings of both studies notwithstanding.

Özkaynak and Thurston [122]

The 1980 total mortality in 98 SMSAs was analyzed, using data from the EPA inhalable particle monitoring network. The independent variables were: % > 65, median age, % nonwhite, log of population density, education, poverty, TSP, suspended sulfates, inhalable particles (IP, diameter < 15 μm) and fine particles (FP, diameter < 2.5 μm). Mean values were reported as (μg/m^3): sulfate, 11; fine particles, 23; total inhalable particles, 48; and TSP, 78. The sulfate measurements that Özkaynak and Thurston used were found to be affected by artifacts from the filters [3]; this is confirmed by the fact that their mean value exceeds those of previous years and other studies.

The authors concluded that the results were "suggestive" of an effect of particles on mortality decreasing with particle size; SO_4^{2-} and FP significant, IP and TSP not. As discussed above, when SMSAs or other large geographic areas are chosen for the observational unit, the finding of increased statistical significance for a small particle variable does not necessarily have health implications, since the effects of large particles will be confined to an area closer to the measuring site and thus will not be detectable by a study using large areas.

In addition, lack of a complete model specification sheds doubt on the validity of the results, since smoking, diet, water hardness, and migration were not accounted for. Contrary to the authors' statement, the confounding effect of smoking cannot be assessed simply by the simple correlation between a smoking index and some air pollution variable. Because of the multicollinearity present, the entire covariance structure must be examined. No other pollutants were evaluated, such as ozone or trace metals. Thus, Özkaynak and Thurston's findings may not be very specific.

As discussed above, the use of SMSAs as the geographic unit of observation may be responsible for the lack of significance of TSP, since central city measurements will not be good estimates of total SMSA exposure. Estimation rather than measurement of IP and FP may contribute enough error to make them less significant than sulfate, which is expected to be less variable over the whole SMSA. The authors reported investigating spatial autocorrelation and regional influences but did not present sufficient results to permit an independent evaluation.

Winkelstein [123]

Winkelstein performed a semi-quantitative study of U.S. lung cancer (LC) and ischaemic heart disease (IHD) mortality for 25 states, from 1968 to 1978. The dependent variables were age-specific lung cancer (LC) and ischaemic heart disease (IHD) rates for white males and females. The independent variable was the prevalence of cigarette smoking, taken from surveys of behavioral risk factors.

The author concluded that smoking appeared to be the common factor linking spatial patterns of lung cancer and heart disease and that the associations revealed in ecological analyses should receive serious consideration. Although air pollution and other environmental factors may be important issues ". . . susceptible to epidemiological investigation . . .", he concluded that they are unlikely to have an important effect on LC or IHD.

The analyses in this paper should be considered exploratory rather than definitive, as was the author's apparent intent. They suggest that smoking could account for even more of the IHD variance than shown directly (25%), since lung cancer "accounted" for 74% of it and smoking is believed to be responsible for 80–90% of lung cancer.

1. The author felt that the 25-state smoking prevalence data was accurate, since it came within 12% of the Tobacco Institute's estimate of national consumption. However, since lung cancer rates were shown to be a better predictor of IHD rates than the indicated smoking prevalence rates, it is possible that these prevalence rates do not give an accurate picture of the actual distribution of cigarette consumption. Reasons could include inaccurate (self) reporting, lags between use of tobacco and resulting illness, varying consumption per smoker, and bias from the omitted states, which include both very low consumption states (Utah) and high consumption states (Nevada).

2. Lung cancer may have a longer lag period than IHD; if smoking time-trends vary spatially, this could confound the interrelationships.

Although the supporting analyses are relatively crude, the author's main findings appear sound. They suggest that ecological studies of mortality patterns must include smoking as a possible explanatory factor. The results also provide support for the use of lung cancer mortality rates as a "marker" for cigarette smoking; i.e., an independent variable that is highly correlated with lung cancer is likely to also be correlated with smoking habits.

Gregor [124]

This is one of the few relatively recent cross-sectional studies employing intra-urban estimates, in this case the census tracts of Allegheny County, PA (Pitts-burgh). The study used five-year average death rates, stratified by age, race, sex, and broad categories of causes (total, "pollution-related" and "non-pollution-related"). Although a large number of independent variables was initially proposed, includ-ing an estimator for smoking based on age and income, the final model employed only education, population density, two climate variables, SO_2, and TSP. The pollution data were interpolated to the census tract level, using a computerized mapping program. Thus, the numbers of degrees of freedom appropriate for testing the significance of these regression coefficients is actually much smaller than the number of census tracts (assumed to be about 200; the figure was not given in the paper). There were from 4 to 14 TSP stations, depending on the year, and 42–47 SO_2 stations (sulfation plates). The interpolation procedure was also used for the climate data; I find it curious to try to explain mortality within a county in terms of climate, since the variability over this scale is small; the climate variables should thus be regarded as "nuisance variables".

Gregor reported statistical significance for TSP but not for SO_2. His coefficients and elasticities for TSP were an order of magnitude higher than has typically been found from interurban studies, but similar to the values reported by Winkelstein et al. for Erie County, NY [89]. His values for SO_2, although not statistically significant, were similar to those found from time-series analysis (12% per mg/m^3). Gregor's TSP results were strongest for the pollution-related causes of death and for the age group 45–64, and were slightly stronger for females. It is difficult to accept a finding that exposure to TSP (which includes large particles not thought to be very harmful to health) could be responsible for over 20% of mortality (as implied by both Gregor's and Winkelstein's results); both studies may be confounded by socioeconomic factors related to the geographic distribution of TSP sources. An alternative explanation could involve the fact that TSP concentrations 10–20 years earlier were much higher (probably by as much as a factor of 2) [108]. It is also possible that the TSP regression coefficients from the interurban regression studies were depressed due to errors in exposure arising from the use of only one or several measurement stations per city.

Jacobs and Langdoc [125]

Studied changes in long-term mortality in parts of Charleston, SC, as a function of both time and place. They identified a portion of the county which was heavily industrialized and which underwent an extensive cleanup from 1968 to 1970. Annual TSP decreased from 228 to 120 μg/m^3 during this period; there were undoubtedly changes in other air pollutants as well, but no data on other species were reported. The authors contrasted the mortality experience of this area (about 38% of the total) to the whole county and to nine adjacent census tracts (about 20% of the total), for which a TSP monitor about four miles away from the industrial area was deemed representative (mean values of 74 and 55 μg/m^3 for 1968 and 1970). Significant changes were noted in deaths from all causes, for heart disease and stroke (all ages and sexes) and for heart disease for both males and females, age 45–64. The authors reported that there were no significant population shifts during this two-year period. However, the spatial comparison neglected possible confounding effects due to racial and socioeconomic differences and possible differences in smoking habits. Since these confounders are unlikely to have changed significantly in the two-year period, I felt that a better comparison would be one taken over time within each detailed area, using the change over time in the entire county as a reference. On this basis, there were substantial decreases in the mortality ratios for both male and female heart disease deaths (age 45–64) and in total deaths from all causes in the polluted area, but not in the comparison area. The slopes of these changes with respect to TSP were similar to the values obtained by Winkelstein [89] and Gregor [124]. The fact that these changes represent increments over time rather than space lends important support to the causal hypothesis, and suggests that pollution affects mortality rather quickly. It is possible that the mortality effects observed in Charleston are acute rather than chronic, and thus it would be interesting to perform a time-series analysis. However, this would be problematic because of the small population involved.

Cross-Sectional Mortality Analyses from Other Countries

Canada

Plagionnakos and Parker [126] studied mortality rates in nine counties in Ontario, from 1976 to 1982. Their model included SO_2, TSP, and sulfate aerosol, as well as variables for smoking, alcohol consumption, medical care, income and education. Their approach pooled cross-sectional and time-series data and employed a dummy variable for secular trend. The counties were selected on the basis of the availability of air monitoring data and included metropolitan areas and industrial areas (Algoma and Sudbury). They analyzed total mortality (excluding injury and poisoning) and respiratory disease mortality, and hospital admissions for all causes and for respiratory causes. They used a logarithmic model, for which the regression coefficients are numerically equal to elasticities.

Air pollution data were obtained from routine monitoring stations, ranging from 2 to 12 per county for SO_2, 2 to 27 for TSP, and 1 to 11 for sulfate. Air quality improved significantly during the period studied. In 1976, annual average SO_2 ranged from 25–65 $\mu g/m^3$; in 1982, 8–30 $\mu g/m^3$. Sulfates were almost constant over this period, ranging from about 8–14 $\mu g/m^3$ by location. TSP data fell into two clearly delineated groupings; the urban areas were in the range 80–95 $\mu g/m^3$ in 1976, while the other counties (including Algoma and Sudbury) ranged from about 45–70 $\mu g/m^3$. By 1982, the two groups had decreased to 65–80 and 40–60 $\mu g/m^3$, respectively. As a result, there was less collinearity in this study than in some previous ones. SO_2 and sulfate were not strongly correlated with socioeconomic variables, and the correlations with TSP were only 0.57 and 0.46, respectively. The correlation over the region between maximum 24-hour and annual average SO_2 was 0.45, but the correlation with alcohol consumption was 0.68, which might make it difficult to deduce any acute effects from SO_2. The maximum 24-hour SO_2 reported was about 800 $\mu g/m^3$. Since some of the pollution sources involved in this study are smelters, it is possible that metal particulates may have contributed.

All four dependent variables were significantly associated with air pollution, with the following elasticities, which are alternatives, not to be summed:

| | Mortality | | Hospital admissions | |
	All causes	Respiratory	All causes	Respiratory
SO_2	0.04–0.06	0.12–0.14	0.06–0.08	0.15–0.18
SO_4^{2-}	0.05	0.10*	0.05	0.18
TSP	—	0.11–0.18	—	—

*p < 0.10

There was a slight tendency towards higher elasticity and significance for 24-hour average SO_2 (as opposed to annual).

These findings appear to be self-consistent and consistent with a causal effect of sulfur oxides (and to a lesser extent, TSP) on long-term measures of health. The results for total mortality are also consistent with findings for 1980 urban mortality in the United States [3].

Great Britain

Early studies of long-term mortality and air pollution in Britain focused on bronchitis and cancer, in relation to the degree of urbanization and various measures of air pollution, some of which were collected over a 10-year period ending in 1954. Pemberton and Goldberg [127] regressed bronchitis mortality rates alternatively against SO_2 (from lead peroxide sulfation candles) and "total solids" (not defined, but assumed to be settleable particulates) and found better correlations mostly with SO_2. The long-term SO_2 levels were in the approximate range 5–120 $\mu g/m^3$. Most of the correlations were significant, although for females they just missed the 0.05 level. They also noted a negative relationship (not significant) between SO_2 and two socioeconomic variables, and thus concluded that confounding of the bronchitis-pollution relationship was not likely. For men, the relationship was stronger in the 45–64 age group; for women, the opposite was true.

Gorham [128] performed a simple regression analysis of bronchitis mortality rates for 53 county and metropolitan boroughs for 1950–4, in relation to sulfate deposits and precipitation pH. He found both to be statistically significant (after logarithmic transformations), the latter with p < 0.001. In his discussion, Gorham felt that the responsible agents might actually be acidic aerosols, a pollutant which is currently receiving attention in the United States.

Daly [129] also studied correlations between bronchitis mortality and air pollution for 1946–50, but was dissatisfied with the available ambient measurement data and devised an index based on the density of consumption of domestic coal (space heating). This procedure produced higher correlations for both males and females (age 45–64), which were not substantially diminished by controlling for socioeconomic conditions. However, none of these early studies considered possible confounding by tobacco smoking. Daly also pointed out that only 18% of the coal consumed in Britain was used for domestic purposes, yet this was the index that was associated with bronchitis mortality. He felt that this was due to poor combustion conditions in domestic fires, which produced much more smoke per unit of mass consumed than industrial combustors (but only proportional amounts of SO_2). For this reason, he felt " . . . it would be premature to attribute the apparently bad effects of air pollution primarily to sulfur dioxide . . . . " In addition, he performed a rudimentary analysis of the impact of sulfur pollution from power stations and concluded that there was no effect on bronchitis mortality.

Stocks [130] looked at the relationships between bronchitis and various cancers (lung, stomach, intestine, breast) and both smoke and "deposit" (settleable particulates) in four different sets of cross-sectional data. The correlation between smoke and "deposit" was just significant; the long-term average smoke levels ranged from 60 to 490 $\mu g/m^3$ (mean = 260 $\mu g/m^3$). He found significant relationships between bronchitis, lung cancer, and stomach cancer and both smoke and deposit, for both sexes, after controlling for population density. Stocks hypothesized that the stomach cancer finding could relate to ". . . exposure of food to dirty air . . ." and pointed out that stomach cancer rates had declined in the United States where the wrapping of food had long been practiced, whereas such wrapping was just beginning in Britain.

Stocks continued his cross-sectional analyses [92] by adding various trace metals and polycyclic organic compounds as independent (air pollution) variables. Most data were from the 1950s and the overall annual average benzopyrene level was 35 ng/m^3. Socioeconomic variables were population density and an index of social class. Although cross-sectional data on smoking habits were not available, Stocks reported from a survey of hospital patients that lung cancer among heavy smokers was not affected by urban residence, but that among light smokers there was a mortality ratio of about 2.5:1 for urban:rural. For 26 locations in Northern England and Wales he found strong correlations between smoke or benzopyrene and mortality due to lung cancer (males), bronchitis, pneumonia, and a weaker correlation with stomach cancer (both sexes). Strong correlations were also found for many of the trace metals. Similar findings were reported when settleable particulates were regressed against mortality for two other cross-sectional data sets, but the relationships were weaker for subareas of London.

In a more detailed analysis, Gardner et al. [131] used measurements of British smoke, SO_2, and Daly's domestic fuel consumption index (ca. 1951) to examine relationships with 1948–54 and 1958–64 deaths from all causes, cardiovascular disease, bronchitis, lung cancer, and stomach cancer. They devoted considerable effort to dealing with possible confounding variables, including drinking water hardness (calcium content), "social conditions", latitude, and rainfall. It is not clear how the latter two factors might be causally linked to mortality, except that the north of Britain tends to be more industrialized. Including these factors in a multiple regression analysis may thus partially account for regional factors, leaving only local variations to be explained by air pollution and "social conditions". The authors noted that these last two factors were highly correlated (r = 0.8) in their data set. Daly's air pollution index was better correlated with all-cause mortality than either smoke or SO_2 for both sexes and thus was used in the multiple regressions. These regressions showed the following, by cause of death:

All causes: Air pollution was highly significant for males aged 45–64 for both periods and for age 65–74 in the earlier period. Social factors were more important for females. Rainfall, water calcium, and latitude were also important.
Cardiovascular disease: Air pollution was just significant for males, ages 45–64. Rainfall, water calcium, and latitude were important for all subgroups.
Bronchitis: Air pollution was significant for six of the eight subgroups. The other factors were mostly not significant.
Lung cancer: Air pollution was significant for seven of the eight subgroups. The other factors were mostly not significant.
Stomach cancer: Air pollution was significant for three of the four male subgroups. The other factors were mostly not significant.

The authors commented on the possible role of cigarette smoking, which they were unable to account for. However, their results do not suggest major confounding since female lung cancer was also associated with air pollution (although females then smoked less) and heart disease (known to be associated with smoking) was not associated with air pollution, for the most part. It is unfortunate that they did not also present multiple regression results for the individual measured air pollutants. We are unable to examine the association between SO_2 and heart disease, for example.

Chinn et al. [132] analyzed mortality in 104 counties and boroughs in England and Wales from 1969 to 1973. The dependent variables were mortality for infants (aged 0–1); males and females aged 45–54, 55–64, and 65–74; for all causes, stomach cancer, lung cancer, breast cancer, influenza, pneumonia, bronchitis, hypertension, ischaemic heart disease, cerebrovascular disease, "other" heart disease, and suicide. The independent variables were (then) current SO_2 and particulates (British smoke), various socioeconomic variables, latitude, temperature, and rainfall. Limited analysis was done using data for water hardness and smoking (limited data) and Daly's index of domestic fuel consumption [129].

The authors concluded: ". . . no consistent relation of smoke or sulfur dioxide with mortality from all causes or with mortality from specified causes postulated *a priori* to be related to pollution. In particular there was no significant association between smoke and mortality rates for respiratory illness. Comparison with results from similar analyses of data from previous decades suggested that a decline in the strength of associations had occurred in parallel with declining levels of the pollution . . . ." The mean values of smoke and SO_2 were 55 and 119 $\mu g/m^3$, respectively.

They found no associations of infant mortality with smoke or SO_2 and some sensitivity in the results to the selection of data sets, especially for all-cause and heart disease mortality. Deleting the London boroughs, which were the highest in SO_2, strengthened some of the SO_2 correlations, especially for females and for deaths from heart disease. Several of these became significant ($p < 0.05$). Including either water hardness or smoking variables weakened the associations of male mortality with SO_2. Even in 1948–50, female mortality from heart disease was not associated with air pollution (as an index variable). Associations were shown more often for the 45–64 group than for ages 65–74.

The importance of this study is its potential to provide an independent validation of the findings of similar U.S. studies. I believe it to be the most comprehensive cross-sectional study of general mortality in Britain based on actual air quality monitoring data. Replications of U.S. studies *per se* cannot provide independent validation because of the strong autocorrelation of both pollution levels and mortality rates (all years tend to resemble each other). The Chinn et al. study provides several interesting insights, although it also has faults.

The strengths of this study included: the geographic areas used were quite small by U.S. standards (maximum 70 square miles) and on average contained 4 monitoring sites each. Thus, exposure to the specific air pollutants considered was characterized much better than in most comparable U.S. studies. The authors report that the intra-borough variation in air quality was relatively modest, and that few monitoring sites were in industrial areas. Both age- and cause-specific deaths were considered, and 5-year averages (1969–73) were used to obtain numerical stability. The data for both SO_2 and smoke contained an adequate range to be able to demonstrate an effect, if it existed. Levels ranged from almost background to almost twice the U.S. standard. SO_2 and smoke were only weakly correlated ($r = 0.25$). Both drinking water hardness and smoking were included, although the smoking analysis was quite crude, since consumption data were available only on a regional basis.

The weaknesses included: only winter pollution levels were used, since summer levels were "uniformly low". If this were strictly true, the degrees of association would not be affected but the (nonstandardized) regression coefficients would be too low. Using year-round data would have provided a more straightforward comparison with other studies. No data were included on frequency of fog, sulfate particles, oxidants, or acid aerosols. No attempt was made to construct long-term average pollution levels from the available historical data. SO_2 was reported to be correlated with male smoking, making it difficult to separate the effects. Latitude was included in the regressions, even though there is no *a priori* reason to suspect an independent effect on health; since latitude was moderately correlated with air pollution ($r=0.5$), its inclusion could result in the underestimation of pollution effects.

Although regressions are presented of current mortality on indices of past pollution levels (back to ca. 1950), no indications are given as to what typical concentrations might have been at that time, or whether the results are consistent with an expected dose-response relationship. The publications give only sketchy details of all the results that must have been available. In particular, only standardized regression coefficients are given (S.R.C. = $B[\text{sigma}_x/\text{sigma}_y]$); the means and standard deviations of the variables are not given. I calculated these statistics for smoke and SO_2 from the tabulated data and concluded that the standard deviations (sigma) could be estimated by (max − min)/6. This was the basis I used to estimate elasticities and ordinary regression coefficients from the information given in the paper.

The findings of this study appear to be in contradiction to many of the U.S. cross-sectional studies, in that it finds almost no effects of small particles (British smoke). The study confirms the importance of considering smoking and drinking water hardness, but provides no confirmation of the hypothesis that the very young and very old are the most susceptible to the effects of air pollution.

The authors downplayed the findings of significant effects of SO_2 on mortality, for the following reasons:

1. The number of significant SO_2 findings was not greatly different from what might be expected due to chance.
2. The authors stated that for SO_2 to be harmful to health in the absence of simultaneous high particle loadings ". . . seems biologically implausible".
3. SO_2 increased in significance when the London boroughs were deleted from the analysis, even though most of the higher SO_2 readings were found there.
4. Since inclusion of crude index of smoking reduced the effect of SO_2 on males, it was felt that use of a better measure of smoking would reduce it still further.

However, the effects of SO_2 on female mortality remained, unaffected by the inclusion of smoking or water hardness, for all causes, hypertension, and bronchitis. In addition, no (positive) air pollution effects were seen for the "null hypothesis" causes of death: breast cancer and suicide. The lack of an effect in London could be due to increased mobility of the population there or possibly to adaptation to previous levels there, which were much higher.

It is curious that ischaemic heart disease failed to show any air pollution effects, although it was the most important cause of death for many of the age-sex groups, and the bi-variate correlations with air pollution were about the same as for all-cause mortality. The report mentioned ambiguities in diagnosis; which may be an explanation. I estimated the elasticities for female all-cause mortality with respect to SO_2 at about 0.04–0.05. A similar value was obtained for all-cause male mortality, without considering smoking. There was insufficient information to make a similar estimate with smoking included. These elasticities are virtually identical to values obtained for all-cause mortality in the United States [3] and Canada [126], and thus I concluded that the study supports the hypothesis that air pollution has an effect on mortality at present concentration levels (in contrast to the authors' conclusions).

Dean et al. [133] studied local variations in mortality and air pollution in Cleveland County (North-East England), from 1963 to 1972. Smoking effects were also emphasized; data on smoking habits of individual decedents were obtained by interviews with relatives, as were qualitative descriptions of occupational exposures. Air pollution exposure was assessed by defining three different residential areas having the following 1972 winter average smoke and SO_2 levels ($\mu g/m^3$): High – 175, 129; Medium – 66, 85; Low – 35, 57. During 1963 to 1972, both smoke and SO_2 levels declined in the "High" area, but only smoke declined in the other areas. For all causes of death examined (lung cancer, chronic bronchitis, coronary heart disease, and cerebrovascular disease for males; lung cancer and chronic bronchitis for females), only the "High" area showed an elevated risk (about 60%, $p < 0.001$), after standardizing for age and smoking. Thus one pollutant cannot be designated as "responsible"; on the basis of smoke, the effect would be about 480% per mg/m^3; on the basis of both smoke and SO_2, about 330% per mg/m^3. These are very large effects, compared to any of the previous studies, but confirm previous findings that dose-response relationships based on local gradients within cities can greatly exceed those based on city-to-city differences. Dean et al. considered several other risk factors singly. For example, with the exception of lung cancer, the effect of social class was comparable to that of air pollution, which raises the possibility of confounding. With the exception of female bronchitis, the effect of air pollution was smaller than that of the lowest smoking category (1–12 cigarettes/day). Some speculative insights may be gained from the effects of occupational exposures (males): exposure to dust had significant effects on lung cancer and bronchitis, but not on heart or cerebrovascular diseases, suggesting SO_2 effects for the latter causes.

In Part 2 of the report, similar findings were shown for three other residential subdivisions, for lung cancer and bronchitis. In addition, the residential distribution of social class was shown and was discounted as a confounding factor.

Japan

Imai et al. [134] combined cross-sectional and time-series analysis of mortality in the area of Hokkaichi, Japan. Trends in age- and cause-specific mortality in a "polluted" area were compared to those in a "clean" area, from 1963 to 1983. The dependent variables were mortality by 20-year age groups for bronchitis and bronchial asthma; the independent variable was SO_2 (measured by sulfation plates). The authors concluded that bronchitis and bronchial asthma mortality responded to improvements in air quality and also showed a corresponding spatial gradient.

This study has only limited application to community air pollution. First, the "polluted" area was also known for the presence of large particle sulfuric acid mist [135] which was probably more irritating than the SO_2 present and was not specifically considered by Imai et al. Second, the specific causes of death involved are not important in the United States (chronic bronchitis deaths = 0.17% of total; bronchial asthma deaths are not listed as such). SO_2 was measured by sulfation plates, which would also be sensitive to sulfuric acid mist, as well as to other environmental factors including wind speed. Thus, coastal areas may have been lower in air concentrations of sulfur oxides than interior areas, for the same sulfation plate reading. According to Kitagawa [134], H_2SO_4 was present in the area from 1965 and at least until 1969, even though control equipment was reportedly installed in 1967. A ratio of 0.48 between SO_3 and SO_2 was reported, which would mean that there may have been as much as 40–60 $\mu g/m^3$ acid in the polluted area (annual average). Such high acid levels do not exist in community air in the United States, for example.

The strongest mortality effects for bronchial asthma were for children, given as the 0–19 age group. Infant mortality usually dominates this age group, and also has a strong socioeconomic component. Since one might expect a lower socioeconomic group to reside in the polluted area, the mortality gradient shown may be only coincidentally associated with the pollution gradient. In addition, the numbers of deaths in each category were quite small; a change in diagnosis of one or two deaths per year could easily change the outcome of the analysis.

Since a court case had been settled in favor of exposed people in this area (1972), there was undoubtedly a high degree of public awareness of respiratory health issues which could have influenced cause-of-death diagnoses at that time. Verification of cause of death by autopsy was not mentioned. Death rates from bronchitis in the polluted area peaked in 1975, and had returned to about the same as 1965 levels by 1977. There was also a strong increase in reported bronchitis deaths in the "clean" area from 1972 to 1975, possibly also because of increased public awareness.

Age adjustment was done on the basis of the 1935 distribution of the Japanese population, which may have been substantially different than that during the period of study, especially given the impact of World War II. Smoking habits were not taken into account in this analysis, nor were deaths given by sex. Thus, it is not possible to judge whether either temporal changes or spatial gradients in smoking habits might have influenced the results. Bronchitis mortality increased dramatically in the "clean" area, especially after 1974 (by a factor of 3 for the over 60 age group), implying the presence of other factors, not accounted for in the analysis.

In spite of what may be an effect on respiratory health in the area, including morbidity effects as reported by Kitagawa, there was no effect on heart disease [136]. Thus, Imai et al.'s findings are inconsistent with those U.S. studies that associate sulfates with increased heart disease mortality (discussed above).

Poland

Mortality rates from random samples of two areas in Cracow were compared after ten years of follow-up (1968–78) [137]. This is one of the few non-ecological cross-sectional mortality studies, since 4355 individuals were studied and their individual characteristics were identified (including smoking habits) and used in the mortality models. The two residential areas were characterized as: high pollution, smoke = 180 $\mu g/m^3$, SO_2 = 114 $\mu g/m^3$; lower pollution, smoke = 109 $\mu g/m^3$ and SO_2 = 53 $\mu g/m^3$. Both of these levels would be considered "high" in comparison to current conditions in the United States. Fifteen percent of the study sample lived in the high pollution area (center of the city). Because of the relatively small numbers of deaths involved, several different models were evaluated, rather than one comprehensive model.

The authors found a marginally statistically significant (positive) effect of air pollution on male mortality (after age adjustment and consideration of confounding variables), but a *negative* effect on female mortality. There was also an interaction between air pollution and smokers for males. The magnitude of the air pollution effect (both smoke and sulfur) was approximately consistent with the findings for Buffalo [89] and Pittsburgh [124] (both intra-urban studies). The counterintuitive (negative) effect for females may have resulted from the small number of deaths in the polluted area [21], lack of interaction with smoking, or differences in exposure due to spending more time indoors. Also, the finding of air pollution effects for men may be an artifact and there may have been a significant pollution effect for either sex.

The validity of the study is supported by the findings of strong effects for smoking (men only), occupational hazards (both sexes) and education (women only). A rural place of birth was beneficial for both men and women. This study should be viewed as providing qualified support for the hypothesis that long-term exposure to air pollution is associated with increased mortality, although it provides no information as to the harmful pollutants. In that sense, the results are similar to those of Morris et al. [138], who compared two nearby towns in Pennsylvania and showed an additive (positive) effect of air pollution and smoking for males (with a tendency towards negative effects for females).

West Germany

Göttinger [139] regressed 1974–8 male and female mortality rates for 41 census tracts in Munich, using both weighted linear and logarithmic models. Air pollutants were TSP, SO_2, and CO, and values for each census tract were obtained by interpolation. The average TSP level was 230 $\mu g/m^3$ and the average SO_2 level, 38 ppt/24 h. Göttinger used two climate variables but did not state why climate should be a variable within a metropolitan area or how many stations were used. In this respect (and in several others) his study resembles Gregor's [124]. He considered using a smoking index variable, but did not since "... inclusion of the index would reduce the air pollution coefficients throughout and in some cases destroy their statistical significance ...." For this reason, the author advises caution in interpreting his results. The socioeconomic variables used were education and population density. He apparently age-adjusted the mortality rates within the broad age groups used (< 45, 45–64, > 64).

Using weighted linear models, Göttinger found statistically significant effects for TSP but not for SO_2 or CO, for both genders, for the upper two age groups, and for all causes, pollution-related causes, and for females for non-pollution-related causes. The magnitudes of the TSP coefficients were remarkably similar to Gregor's findings and reasonably consistent with the findings from other intra-urban studies. Göttinger used logarithmic models to estimate elasticities, but reports values much lower than would be obtained directly from the linear models. For example, he reports a value of 2.7% for male mortality from all causes, age 45–64, whereas I estimated about 26%, based on the linear model coefficient. For comparison, Gregor [124] reports an elasticity value of 40%.

Concluding Discussion

Physiological Considerations and Trends for Specific Diseases

Respiratory Diseases

The physiological connection between air pollution and respiratory disease appears to be straightforward, since the lung is the primary target organ for inhaled material. Many of the early studies focused on respiratory diseases, especially in Britain, presumably for this reason. In addition, respiratory deaths have long been more common in Britain than in the United States where these causes of death constitute such a small fraction of the total (< 2%) that they are difficult to study. However, asthma deaths have been increasing in recent years [140] and deaths from asthma, bronchitis, and especially emphysema have been increasing since 1950 even faster than lung cancer [141]. The most likely cause is smoking, especially since particulate air pollution has declined during this period.

Heart Disease

In some recent studies, strong associations between air pollution and mortality have been found for diseases of the circulatory system. The physiological hypothesis is based on the increased cardio-respiratory burden of breathing polluted air, which is qualitatively substantiated by the findings of impaired lung function as an independent risk factor for heart disease mortality [142]. Statistically, the associations may also be related to the fact that cardiovascular causes of death are the most common among persons over 45 years of age.

One possible source of confirmation of a causal association between air pollution and heart disease could be the trends over space and time in the past, when U.S. cities were generally much more polluted than they are now [108, 143]. This air pollution was primarily sulfur oxides and particulate matter derived from uncontrolled coal combustion, much of it for home heating. The year 1940 was close to the peak for urban air pollution in many cities, as inferred from data on fuel use data and the few available air quality measurements [108, 143]. Gover [144] analyzed U.S. heart disease trends from 1920 to 1940, by geographic area and size of city. Because of concomitant trends in smoking (which then was much more common among males), the data for female mortality are probably the most useful for this purpose. For the entire United States, mortality rates increased about 40% from rural to urban (greater than 100,000 population) areas for age-adjusted female "heart disease", and about 20% for "all cardiovascular renal diseases". Comparisons with more recent data are difficult because of changes in procedures for coding the causes of death [145]. An additional complicating factor during this period is the effect of changing competing risks due to influenza, pneumonia, and tuberculosis.

By region, the highest rates for mortality in urban areas in 1940 were in the Middle Atlantic, New England, and East North Central regions; within the Middle Atlantic region, the urban-rural gradient was inconsistent, especially for "all cardiovascular renal diseases". Although the time trends from 1920 to 1940 for mortality from "all cardiovascular renal diseases" and the inferred SO_2 concentrations were reasonably consistent in the most polluted and highest mortality regions (approximately flat), the trend for "heart disease" was strongly upward. Also, the gradient in inferred urban-rural air pollution (SO_2) was much stronger than the mortality gradient. All-in-all, I concluded that Gover's data provide only qualified support for a weak effect of urban air pollution on female heart disease mortality.

Cardiovascular mortality patterns for 1950 were analyzed by Enterline and Stewart [146] (state level) and by Enterline et al. [147] (metropolitan areas). Unfortunately, the latter analyzed male rates only, which are thought to be confounded by differences in smoking habits. The 1950 geographic pattern for age-adjusted mortality in white females due to coronary heart disease shows a range of about a factor of two, with the highest rates located in the Northeast (Delaware and Pennsylvania and all states to the north), Illinois, California, and Nevada. With the exception of Nevada and rural New England, these locations had high levels of urban air pollution. Note that TSP in Los Angeles in the early 1950s was higher than in New York and only slightly lower than in Chicago [108]. The authors offered no causal explanations for the patterns but noted that the geographic patterns were similar for both males and females, and for different levels of urbanization. To explain regional differences in rural mortality rates in terms of air pollution, it would probably be necessary to consider long-range transport, which would tend to favor sulfates over TSP as a contributing factor. Note the similarity in geographic pattern between Figs. 19 and 27, for example.

Confounding by smoking habits is also suggested by the analysis of Sauer et al. [148] for the southeastern states, which contained areas of both high and low rates of cardiovascular mortality in 1950. The authors noted that the correlation between lung cancer and coronary heart disease in Georgia was 0.94, but did not consider the possibility of smoking as a causal factor for both diseases. A parallel analysis for females was not presented, but high correlations were shown between men and women in the same age group.

Confounding by migration was explored by Hechter and Borhani [145] by contrasting 1950 and 1960 mortality rates in California by location of birth. Interest in this problem had developed by noting that heart disease deaths in California declined from 1950 to 1960 while increasing in the United States as a whole, and that the California population increased by about 50% during this period. Hechter and Borhani's analysis also shows age-adjusted deaths for white females due to arteriosclerotic heart disease (ASHD) deaths above the national average for both periods for the Middle Atlantic, New England, the northern portion of the South Atlantic, and the East North Central regions, plus Louisiana, California, and Nevada (listed in decreasing order). Within California, female ASHD mortality rates were only slightly affected by urbanization. ASHD mortality rates were higher for both males and females for decedents born in states other than California. Foreign-born males had lower rates than the U.S. born while foreign-born females had higher rates. However, recent migrants (within the previous five years) had generally lower or similar rates than native-born Californians, and when recent migrants were excluded, the overall trend from 1950 to 1960 was still downward. Sauer and Donnell [149] noted that migrants to Florida from the north tended to have lower mortality rates, confirming that the healthier portion of the population is more likely to move to an advantageous area. The finding that native-born Californians had ASHD rates lower than non-recent migrants may reflect socioeconomic differences arising from the migrations during the 1930s.

The 1960 geographic patterns among State Economic Areas (groups of counties) for various causes of death are presented by Sauer and Brand [150], who noted similarities with the 1950 patterns. By disease, they noted similar patterns for males and females and for cancers, coronary heart disease, and cardiovascular-renal diseases. Rheumatic heart disease, chronic nonspecific respiratory diseases and accidental/violent deaths all had independent patterns. The respiratory disease pattern for white males (ages 35–74) did not implicate major metropolitan areas, but instead, areas in the intermountain West and in Appalachia. The correlations among state economic areas presented by Erhardt and Berlin [151] for lung cancer versus cardiovascular-renal disease (0.55 for white males, all areas; 0.21 for white females, all areas; 0.35 for metropolitan white males; 0.50 for non-metropolitan white males) suggest that smoking may be a common factor, but that other metropolitan factors are also important. Air pollution could be one of these factors.

Kitagawa and Hauser [152] performed a correlation analysis of 1960 SMSA mortality (all causes) and found an effect due to TSP of about the same magnitude as Lave and Seskin [101], after controlling for race, sex, education, population density, manufacturing employment, income, housing conditions and climate.

They mentioned (but did not identify) one other air pollution index, which was apparently not significant.

Cancer

At current community concentration levels, air pollution is thought to be a minor contributor to the incidence of cancer, including respiratory cancer [153]. Extreme levels could contribute to earlier demise of terminally ill patients, however.

Cigarette smoking is overwhelmingly the cause of lung cancer, as determined by many studies [86]. The similarity of lung cancer rates among nonsmokers in many locations of greatly different environmental conditions (United States, United Kingdom, and Finland, for example [153]) is the possibly the best evidence of lack of a strong effect due to air pollution. In addition, if a linear dose-response relationship is assumed, the relative doses of carcinogenic compounds received via smoking and via community air imply only small effects due to the latter [108].

Several of the early studies reviewed above found associations between particulate air pollution and mortality from stomach cancer, which has been declining throughout much of the world since at least 1950 [153]. The finding of very similar rates of change in both "polluted" and "unpolluted" countries [153] argues against a strong (widespread) environmental component, although it is difficult to rule out local contributions to specific populations from specific pollution sources.

Summary of Long-Term Effects of Air Pollution on Mortality

Previous Summaries of Cross-Sectional Studies

There have been many previous critiques and summaries of air pollution-mortality studies [4–8, 102–105, 111, 154]. However, few of these adequately considered the effects of errors in estimating air pollution exposures, and there are now some new results that were not available to previous authors. For example, Thibodeau et al. [111] reanalyzed the Lave and Seskin data set [101] using the same basic data and variables, but correcting for errors in the data and adding some statistical techniques. This type of reanalysis provides confidence in the results of the original authors with respect to the absence of numerical errors, but cannot provide insight into robustness with respect to choices of variables and observations, which can be a more important issue.

Ware et al. [154] described the types of studies reviewed in this chapter as "observational", since they do not involve experiments on human subjects. With respect to the long-term studies, Ware et al. concluded that, ". . . The model can only be approximately correct, the surrogate explanatory variables can never lead to an adequate adjusted analysis, and it is impossible to separate associations of mortality rate with pollutant and confounding variables. This group of studies, in our opinion, provides no reliable evidence for assessing the health effects of sulfur dioxide and particulates . . . ." However, as discussed below, they did not support this conclusion by attempting to explain the differences among study results.

Ricci and Wyzga [7] emphasized statistical issues and proposed a list of issues for further research but arrived at no overall conclusion regarding the validity of long-term effects of air pollution on mortality.

Evans et al. [6] critically reviewed some of the published cross-sectional mortality studies, including a quantitative comparison of regression coefficients and a reanalysis of the Lave-Seskin 1960 data with new variables added. For this reanalysis, the dependent variable was total 1960 mortality for 66 to 98 SMSAs. The independent variables included $\% > 65$, median age, % nonwhite, log of population density, % college, smoking index, % poverty and mean sulfate concentration. Other variables explored included occupational, home heating, climate and housing variables; mean values for TSP, iron, manganese, and B(a)P. In general, the sulfate variable lost statistical significance in this reanalysis (in part because the mean rather than minimum concentrations were used; see the discussion of Lave and Seskin, above [101]); the elasticity with respect to total mortality was about 0.03. None of the other pollutants provided any better results.

Evans et al. concluded ". . . important questions related to the interpretation of cross-sectional studies remain unanswered. For example, it is unclear which specific pollutants are responsible for any observed effect; the shape of the exposure-response relationship is unresolved; and it is possible that the observed effects are due, in part, to confounding or systematic misclassification . . . ." Nevertheless, the authors were of the opinion that ". . . the cross-sectional studies reflect a causal relationship between exposure to airborne particles and premature mortality . . . ."

My 1985 review [8] covered both short and long-term studies and discussed the interpretation of numerical differences in coefficients. I was confident that the short-term studies reflected causal relationships, but held that it was not possible to determine quantitative relationships with confidence. In our analysis of 1980 mortality rates in U.S. cities [3], we tried to approach the causality issue by comparing the ecological regression coefficients for non-pollutant variables (such as smoking, poverty and demography) to previous estimates based on individuals. The degree of success of such comparisons would then provide support for a causal interpretation of the pollution effects.

In the United States, health professionals seldom mention environmental pollution factors in discussing current mortality patterns and rates of change. For example, for heart disease, smoking, blood pressure and levels of serum cholesterol are thought to be the primary risk factors. However, Kleinman et al. [155] attempted to reconcile the observed differentials in regional and urban-suburban rates of coronary heart disease (CHD) mortality in terms of the corresponding differentials in these risk factors. They concluded: ". . . differentials in these risk factors cannot account for the lower CHD death rates observed in the West as compared to other regions or among suburban as compared to urban residents . . . ." Thus, this finding supports an air pollution risk factor, in addition to those listed above.

Meta-Analysis

A meta-analysis treats the results from individual studies as "observations" in an integrative analysis of the group of studies as a whole [156]. The purpose of meta-analysis is to provide confidence in the existence or plausible magnitude of an effect. Evans et al. [6] attempted a quantitative meta-analysis by grouping results from various studies and compiling frequency distributions of regression coefficients. They did not attempt to explain the differences in coefficients in terms of the overall designs of the various studies, and did not explicitly consider which "observations" were independent and which were essentially replications. They acknowledged that their meta-analysis approach did not ". . . improve the state of knowledge concerning the true damage coefficients . . . ."

In addition, a meta-analysis must consider the issue of publication bias: research has shown that studies with positive findings are more likely to be accepted for publication than findings of "no effect" [156]. Thus, having included unpublished reports (the "gray" literature) in a review such as this should help provide balance, provided judgment was exercised with regard to the designs of studies and the validity of findings.

In a meta-analysis, the characteristics of a study may serve as explanatory variables with regard to differences in findings. Table 6 summarizes the major long-term mortality studies reviewed in this chapter. The study characteristics to consider include the degree of age-sex-race control, the adequacy of control for socioeconomic effects (including lifestyle variables such as smoking), the range of variability in air pollution, and the adequacy of estimation of exposure, in addition to the times and places studied. Many of these factors are somewhat subjective, and thus I have not attempted to use them in a formal quantitative meta-analysis.

Some conclusions arise from considering Table 6 (some of which have been discussed above in more detail):

1. The indicated effects of air pollution on mortality are sensitive to the data set analyzed and to the socioeconomic variables used to explain non-pollution effects.
2. Respiratory deaths tend to be associated with particulate air pollution, especially in the pre-1970 period when particulate concentrations were higher. Cardiovascular mortality may be associated with sulfur oxides air pollution.
3. During the mid-1970s in the United States, the association between sulfur oxides and mortality became substantially stronger, for unknown reasons.
4. The association between mortality and particulate air pollution has become less significant in both the United States and Britain. This may be linked to a reduction in ambient concentrations resulting from air pollution control efforts.
5. The association between air pollution and mortality tends to be substantially stronger within urban areas than between urban areas. It is not clear how much of this results from better characterization of exposure to air pollution and how much from sharper gradients in personal lifestyle characteristics.
6. As better account is taken of confounding variables, associations between air pollution and those causes of death without physiological links with air pollution (such as "all cancers") tend to lose significance.
7. There is still ambiguity as to which pollutants are associated with all-cause mortality. It is possible that this reflects different specific pollutant-disease relationships.

The above considerations relate primarily to the *existence* of associations between long-term exposure to air pollution and mortality. Table 6 provides no

Table 6. Comparison of Cross-Sectional Studies

Authors (footnotes)	Time period	Locations	Age-sex-race control[a]	Socio-economic variables[b]	Pollutant ranges[c]	Exposure accuracy[d]	Significant pollutants-diseases[e]	
Mills, Mills-Porter [83]	1929–46	5 cities	sex	none	TSP, 100–700	poor	dust	Resp. dis.
Mills [84]		Chicago	age-sex	none	TSP, 100–500	poor	dust ⎱ SO_2	Resp. dis.
Stocks [90]	1950–4	N. England & Wales	sex (age-adj)	pop. density, social class	smoke, 15–562; BP, 0.001–0.1 (other H/C, tr. metals) settl. dust	OK / ? / ?	smoke / smoke / smoke / dust / dust / dust	Lung cancer / Resp. dis. / Stomach ca. / Lung cancer / Bronchitis / Stomach ca.
Winkelstein et al. [89, 90], [91]	1959–61	Buffalo	a-s-r	income, educ., occupations	TSP, 75–206; SO_2, 61–385	good / ?	TSP ⎱ SO_2	All causes / ASHD / Stroke age 50–69 / CRD M, F / Stomach Ca.
Zeidberg et al. [97–100]	1949–60	Nashville	a-s-r	income, educ., housing	COH, <0.35 –> 1.1; SO_2, <13 –> 34	poor / poor	TSP / soiling ⎱ SO_2	All CV / Stomach Ca. / Total Resp.
Lave & Seskin [101]	1960	US SMSAs	a-s-r	pop. dens., poverty, pop; others *= significant w/add'l vars.	TSP, 40–224; SO_4, 3–28	poor / poor	TSP* + SO_4 / TSP + SO_4* / TSP* / SO_4 / SO_4	All causes / All CV Dis. / Resp. Dis. / All Cancers / Stomach Ca.
	1969	US SMSAs	a-s-r	pop. dens., poverty, population	TSP, 29–188; SO_4, 2–19	poor / OK	TSP + SO_4	All causes
Chappie & Lave [107]	1974	US SMSAs	age +, race +	pop. dens., income, pop., education, smoking, alcohol use occupation mix	TSP, 75; SO_4, 9.6	poor / OK	SO_4	All disease causes
Lipfert [108]	1959	US cities	age +, race +	pop. dens., poverty, housing, smoking, education	TSP, 40–224; SO_4, 3–28; Fe, 0.5–13; Mn, 0.01–2.2	OK / OK / poor / ?	Mn	All causes

Table 6. (*Continued*)

Authors (footnotes)	Time period	Locations	Age-sex-race control[a]	Socio-economic variables[b]	Pollutant ranges[c]	Exposure accuracy[d]	Significant pollutants-diseases[e]	
Lipfert [108]	1969–71	US cities	age, race +	pop. dens., poverty, housing, smoking, educ., birth rate	TSP, 29–188 SO$_4$, 2–19 Fe, 0.2–9.5 Mn, 0.01–2.0	OK OK poor ?	TSP + Mn Mn Mn Fe TSP + Mn	All causes All cancers Resp. Ca. Resp. Dis. Other, incl. CV
Lipfert [80]	1970	US SMSAs	age, race +	poverty, migration, pop. dens., pop., diet, smoking, water hard., home heat fuel use	TSP, 29–188 SO$_4$, 2–19 Fe, 0.2–9.5 Mn, 0.01–2 ozone, 30	OK poor poor poor poor	TSP TSP Mn*	{ All causes { Males, < 65 { Females, < 65 { Males, > 65
Crocker et al. [109]	1970	US cities	age + , race +	income, education, smoking, diet, medical care, climate	TSP, 102 SO$_2$, 27 NO$_2$, 143	OK poor poor	(none) TSP SO$_2$	All causes Pneum., flu infant
Mendelsoh & Orcutt [115]	1970	US County Groups	a-s-r	income, education, mar. status, migration, pop. dens., housing, region	TSP, 70 SO$_4$, 10.5 NO$_3$, 3.0 SO$_2$, 21 NO$_2$, 50 ozone, 28 CO, 4600	poor OK poor poor poor poor poor	SO$_4$	All diseases
Namekata et al. [120]	1971–5	Chicago (76 areas)	age-adj.	income, housing, education	TSP, 83 SO$_2$, 41	good good	TSP TSP TSP	Heart dis. Liver cirh. Emphys.
Özkaynak & Thurston [121]	1980	US SMSAs	age + , race +	pop. dens., poverty, education	TSP, 78 SO$_4$, 11 Inhal. part., 48 Fine part., 22	poor poor poor OK	FP SO$_4$	All causes All causes
Gregor [123]	1968–72	Allegheny Co. (cens. tracts)	a-s-r	education, pop. dens., climate	TSP, 122 SO$_2$, ?	good good	TSP TSP	All causes "poll. rel." causes

Study	Years	Location	Adjustment	Independent variables	Mean values	Quality	Pollutant	Cause
Jacobs & Langdoc [124]	1968–70	Charleston, SC (two areas)	age-sex	none	TSP, 55–228	good	TSP / TSP / TSP	All causes / Heart dis. / Stroke
Plagionnakos & Parker [125]	1976–82	Ontario (9 counties)	age +	income, education, alcohol, smoking, medical care	TSP, 40–95; SO$_2$, 8–65; SO$_4$, 8–14	good / good / good	TSP / {SO$_2$ / SO$_4$}	Resp. Mort. / Total Mort.
Chinn et al. [127]	1969–73	UK boroughs	age-sex	pop. dens., housing, social class, climate smoking, water hard.	smoke, 15–225; SO$_2$, 20–320	good / good	SO$_2$ / SO$_2$ / SO$_2$	All causes / Hypertension / Bronchitis
Dean et al. [129]	1963–72	Cleveland Co., (3 areas)	sex (age-adj)	smoking	smoke, 35–175; SO$_2$, 57–129	good / good	{Smoke / SO$_2$}	Lung Cancer / Bronchitis / male CHD / Stroke
Krzyzanowski & Wojtyniak [133]	1968–78	Cracow, Poland (2 areas)	sex (age-adj)	smoking, occupational exposure, birthplace education	smoke, 109–180; SO$_2$, 53–114	good / good	{Smoke / SO$_2$}	All causes
Göttinger [135]	1974–8	Munich, FRG (cens. tracts)	age-sex	education, density, climate	TSP; SO$_2$; CO	good	TSP	All causes
Lipfert et al. [111]	1980	US cities	age +, race +	pop. chng., poverty, smoking, migration water hard., birth rt	SO$_4$, 1–15; SO$_2$, 0.3–46; NO$_x$, 0.7–51; Mn, 0.02–0.2; Fe, 0.2–4.3; ozone, 37–200; CO 2250; Inh. part. 40; Fine part. 20; TSP	good / good / good / ? / Fe / poor / poor / OK / good / OK	{SO$_4$ / SO$_2$ / NO$_x$ / Mn}	All causes

[a] Age-sex indicates separate regressions by age and sex. a-s-r = age-sex-race. Age + (race +) indicates that a population age (race) descriptor variable was used as an independent variable in the regressions.

[b] Independent variables listed are consistent with the regression results shown; many authors ran regressions for various deletions and combinations from the list.

[c] Mean values are given when ranges were not available. Units are $\mu g/cu.m.$

[d] Subjective judgement involving numbers of pollution monitors and temporal coincidence of pollution and mortality.

[e] TSP + SO$_4$ indicates both pollutants were significant in a joint regression. Brackets indicate it was not possible to distinguish among the pollutants in the group.

guidance as to the magnitude of effects or whether they are related to current or previous exposures.

The available studies of lag effects [3, 101, 108] are all inconclusive and relate primarily to the effects of sulfur oxides. Other evidence supports the conclusion that lags are relatively short (< 10 years), if they exist at all: the findings of Jacobs and Langdoc in Charleston, SC [125] imply an immediate (same year) response; Plagionnakos and Parker [126] got a very good fit to their pooled cross-sectional time-series data without using lags. The evidence on lags is less clear with respect to TSP: TSP apparently ceased to be a significant mortality factor in the mid-1970s in the United States, and it cannot be determined if this relates to the change from 1969 to 1974 (national average levels of 86 versus 75 $\mu g/m^3$) or to the change since 1959 or earlier (112 versus 75 $\mu g/m^3$). Data from Britain [132] do not support an extensive lag period for smoke pollution, nor do the Winkelstein studies in Buffalo, NY [89–96]. However, TSP effects from U.S. inter-urban studies may be badly compromised by insufficient numbers of monitoring stations and the resulting poor characterization of population exposures.

The magnitude of air pollution effects on total mortality was estimated at about 5% in three separate studies (all of which controlled for smoking): U.S. cities [3], Ontario counties [126] and British boroughs (females only) [132]. This agreement among independent data sets provides the best evidence that such associations are not likely to be due to data artifacts. Several independent studies found substantially stronger effects within cities or urban areas (as much as an order of magnitude higher): Buffalo [89], Pittsburgh [124], Cleveland County (U.K.) [133], Cracow [137] and Munich [139]. Also, the Cracow study dealt with individuals and thus was not "ecological"; however, it also found negative associations between female mortality and air pollution.

Additional Causal Issues Regarding Long-Term Effects

The above summary presents a relatively consistent picture which could lead to the acceptance of the cross-sectional associations as causal. However, there are other considerations and some inconsistencies as well.

Within-City Findings

The finding of relatively consistent effects within cities in four different countries seems to provide strong evidence for causal relationships. Yet it is possible that *all* the studies may be confounded by the same phenomenon, which could transcend national boundaries: the desire of those who are able to do so to live in a clean environment. The interaction between economic and environmental factors was clear for many causes of death in the Erie County studies (Figs. 21–26). Further accounting for personal lifestyle factors accompanying the economic gradients could modify the apparent pollution relationships. There is no *a priori* reason to expect that these factors would operate differently in a different country with the same level of industrial development.

Trends in the Sulfur Oxides Associations

The most puzzling finding and a possible inconsistency is the apparent emergence in the United States in the mid-1970s of sulfur oxides as a significant factor, after controlling for many socioeconomic factors [3, 107, 116]. During this time, air quality was improving in general, although the changes in measured sulfate concentrations since the late 1950s have been modest. In many U.S. cities, SO_2 levels declined substantially, beginning about 1970. In addition, this was a period of decline in heart disease mortality in almost all regions and segments of the population [157]. Several hypotheses come to mind:

1. The indicated SO_x associations may be an artifact of the analysis and thus may represent some undefined regional lifestyle factors.
2. The SO_x associations may have been present before the mid-1970s but were then masked by other risk factors for heart disease. When the other risk factors declined in relative importance as a result of changes in lifestyles, the effects of air pollution became apparent.
3. The SO_x associations may have previously been masked by the effects of TSP. When TSP concentrations declined and lost importance, the effects of SO_x became apparent.
4. The mortality analyses before the mid-1970s may have been flawed and incapable of detecting the effects that were present.
5. The chemical composition of sulfate aerosol (acidity, for example) may have changed since the early 1970s and become more harmful to health.

While there is no evidence to support the first hypothesis (see Ref. [3], which shows effects *within* regions) or the last one, the other three cannot be ruled out. Resolving them would add consistency to the meta-analysis, but will not address the issue of causality *per se*.

Support from Findings on Passive Smoking Effects

A recent interesting development is the growing evidence of the harmful effects of passive smoking (breathing environmental tobacco smoke). In terms of community air pollution, the species involved are fine particles, carbon monoxide, and oxides of nitrogen (sulfur oxides are not involved). Although these findings are still debated, they include heart disease [158, 159], and Wells [160] has estimated that about 2.5% excess deaths may result, mostly from heart disease. For comparison with community air pollution levels, tests in the Netherlands [161] showed that average indoor fine particle concentrations increased by about 30–90 $\mu g/m^3$ per smoker in the household. This finding suggests a dose-response relationship similar to those obtained from some of the cross-sectional studies of community air pollution, but provides no support for associations between sulfur oxides and mortality. Possible passive smoking effects should also signal caution in interpreting the results for females as being free from tobacco interference.

Ecologic Regression Considerations

The primary objection to ecologic regression relates to the lack of specificity of the affected individuals and the exposed individuals, when groups are used in the regression analysis. This objection is most valid when the pollutant is very localized (such as emissions from a toxic waste dump) or when the disease is relatively rare (such as leukemia). However, this objection becomes de minimus for regional pollutants, such as fine particles or sulfates, and for mortality from all causes or from very common causes (such as heart disease). Furthermore, one study reviewed in this section [137] dealt directly with individuals. Thus, it may be argued that the fundamental problems of ecologic regression should not be viewed as an insurmountable problem for cross-sectional studies.

Ecological studies are often viewed as suitable only for generating hypotheses, to be confirmed by more definitive controlled studies. Several such hypotheses have been generated by this review:

1. Any thresholds of no effect must lie below the concentration levels experienced in the studies reviewed. (This statement must be viewed in the context of a large population containing individuals having a range of susceptibilities.)
2. In a large population, different people will be sensitive to different pollutants, with respect to all-cause mortality.
3. Respiratory causes of death tend to be more associated with particulate air pollution than with sulfur oxides.
4. Cardiovascular causes of death tend to be more associated with sulfur oxides than with particulates.
5. The long-term effect on mortality is only slightly greater than the annual sum of short-term effects.
6. The short-term mortality response lags the air pollution dose by a few days, at most.

Conclusions

This chapter reviewed the associations between air pollution and mortality for both short-term and long-term responses. New results continue to be published, and it is probably safe to say that the definitive study of air pollution and mortality has not yet appeared. On the whole, the evidence for short-term effects is more convincing, primarily because of the severe air pollution episodes of years past. In addition, I have shown that the results of time-series analyses of daily mortality form a self-consistent continuum that it is consistent with the effects of the episodes. However, it is not possible to distinguish the effects of individual pollutants, and there is evidence that both particulates (smoke) and SO_2 are important, especially in combination. Other air pollutants may contribute and there may be interactions with weather factors. The effect of the duration of short-term pollution episodes is largely unquantified.

The long-term studies are also seen to be reasonably self-consistent when viewed as a whole. The results of any one study can be strongly affected by the data set used and the degree to which potential confounding variables are managed. Three recent studies in different countries each find an effect of about 5% on total mortality, but, as with the time-series studies, it is not possible to determine the responsible pollutant(s) with certainty. (For this reason, I tried to deemphasize the numerical values of long-term dose-response functions in this review). Some air pollutants, such as carbon monoxide, may have not been associated with mortality, in part because population exposures were poorly measured. Some pollutants may be confounded with socioeconomic factors.

In general, the magnitudes of these findings suggest that the long-term effects on mortality may exceed the short-term effects in terms of the percentages of excess deaths. At the least, the findings of the long-term studies suggest that short-term effects may persist at the relatively low concentrations which are currently experienced in many locations.

Since data on the health effects of air pollution are frequently used to support arguments for increased pollution abatement, future studies on this issue should include longitudinal studies of the long-term effects of previous abatement efforts. In addition, the results of "observational" studies should be used as a basis for confirmatory laboratory and clinical research studies.

Acknowledgments. This review has benefited from many discussions with colleagues at Brookhaven National Laboratory and elsewhere. The data and previous support provided by the Electric Power Research Institute are acknowledged. The manuscript was edited by Avril Woodhead.

References

1. McDonald JC, Drinker P, Gordon JE (1951) Am. J. Med. Sci. 221:325
2. Leamer EE (1978) Specification Searches, Ad Hoc Inference with Non-experimental Data. Wiley, New York
3. Lipfert FW, Malone RG, Daum ML, Mendell NR, Yang CC (1988) A Statistical Study of the Macroepidemiology of Air Pollution and Total Mortality, BNL Report 52122, to U.S. Dept. of Energy
4. Lipfert FW (1978) "The Association of Human Mortality with Air Pollution: Statistical Analyses by Region, by Age, and by Cause of Death", Ph.D. Dissertation, Union Graduate School, Cincinnati, Ohio. (available from Eureka Publications, Mantua, NJ 08051) Also see Lipfert FW, "The Association of Air Pollution and Human Mortality: A Review of Previous Studies", APCA Paper 44.1, presented at the 70th APCA Annual Meeting, Toronto
5. International Electric Research Exchange (IERE) (1981) Effects of SO_2 and Its Derivatives on Health and Ecology, Vol 1, Human Health, (available from EPRI Research Reports Center, P.O. Box 50490, Palo Alto, CA 94303)
6. Evans JS, Tosteson T, Kinney PL (1984) Env. Int. 10:55
7. Ricci PF, Wyzga RE (1983) Env. Int. 9:177
8. Lipfert FW (1985) Env. Sci. Tech. 19:764
9. Mazumdar S, Schimmel H, Higgins I (1980) Relation of Air Pollution to Mortality, An Exploration Using Daily Data for 14 London Winters, 1958–72, report prepared for the Electric Power Research Institute, Palo Alto, CA

10. Firket J (1936) Trans. Faraday Soc. 32:1192
11. Roholm K (1937) J. Ind. Hygiene Toxic. 19:126
12. Schrenk HH, Heimann H, Clayton GD, Gafafer WM, Wexler H (1949) U.S. Public Health Bull. 306, Washington, DC
13. Mills CA (1954) Air Pollution and Community Health. Christopher Publishing, Boston
14. Ciocco A, Thompson DJ (1961) Am. J. Public Health 51:155
15. Logan WPD (1953) Lancet 264:336
16. Brasser LJ, Joosting PE, Van Zuilen D (1967) "Sulphur Dioxide—to what level is it acceptable?" Report G300, Dutch Research Institute for Public Health Engineering
17. Ball DJ, Hume R (1977) Atm. Env. 11:1065
18. Lipfert FW (1988) Changes in Exposure to Air Pollution in Geater London, 1954–56. Memorandum prepared for the Electric Power Research Institute
19. Holland WW, Bennett AE, Cameron IR, Florey CduV, Leeder SR, Schilling RSF, Swan AV, Waller RE (1979) Am. J. Epidemiology 111:525
20. McCarroll J, Bradley W (1966) Amer. J. Public Health 56:1933
21. Greenburg L, Jacobs MB, Drolette BM, Field F, Braverman MM (1962) Public Health Reports 77:7
22. Loewenstein JC, Bourdel MC, Bertin M (1983) Rev. Epidem. et Sante Publique 31:163
23. Riggan WB, VanBruggen JB, Truppi LE, Hertz M (1976) "Mortality models: a policy tool", presented at the Conference on Environmental Modeling and Simulation, Cincinnati, Ohio
24. Stebbings JH Jr, Fogleman DG (1979) Amer. J. Epidemiology 110:27
25. Wichmann HE et al. (1988) Health Effects During a Smog-Episode in West Germany in 1985. Env. Health Perspectives (in press)
26. Bull GM (1973) Br. J. Prev. Soc. Med. 27:108
27. Rogot E, Blackwelder WC (1970) Public Health Rep. 85:25
28. Hodge WT, Nicodemus ML (1980) Human Biometeorology: A Selected Bibliography, NOAA Technical Memorandum EDIS NCC-4
29. Russell WT (1926) Lancet 2:1128
30. Logan WPD (1949) Lancet, January 8, 78
31. Gore AT, Shaddick CW (1958) Br. J. Prev. Soc. Med. 12:104
32. Martin AE, Bradley WH (1960) Mon. Bull. Minist. Health Service Lab. Serv. 19:56
33. Martin AE (1964) Proc. Roy. Soc. Med. 57:969
34. Commins BT, Waller RE (1967) Atm. Env. 1:49
35. Waller RE, Commins BT, Lawther PJ (1965) Brit. J. Ind. Med. 22:128
36. Macfarlane A (1977) Brit. J. Prev. Soc. Med. 31:34
37. Mazumdar S, Schimmel H, Higgins I (1982) Arch. Env. Health 37:213
38. Goldstein IF, Landau E, Van Ryzin J (1983) Arch. Env. Health 38:122
39. Mazumdar S, Schimmel H, Higgins I (1983) Arch. Env. Health 38:123
40. Ostro B (1984) Env. Health Persp. 58:397
41. Schwartz J, Marcus AH (1966) Statistical Reanalyses of Data Relating Mortality to Air Pollution During London Winters 1958–72, U.S. Environmental Protection Agency, Washington, DC
42. Shumway RH, Tai RY, Tai LP, Pawitan Y (1983) Statistical analysis of daily London mortality and associated weather and pollution effects, California Air Resources Board, Sacramento, CA
43. Roth HD, Wyzga RE, Hayter AJ (1986) Methods and Problems in Estimating Health Risks from Particulates, in: Aerosols: Research, Risk Assessment, and Control Strategies, Lee SD, Schneider T, Grant LD, Verkerk PJ (eds) Lewis Publ, Inc., Chelsea, Michigan, p 1047
44. Ellison JMcK, Waller TE (1978) Env. Research 16:302
45. Greenburg L, Jacobs MB, Drolette BM, Field F, Braverman MM (1962) Pub. Hlth. Rep. 77:7
46. Shils ME, Skolnick JN (eds) (1978) Bull. NY Acad. Med. 54:983
47. Hodgson TA Jr (1970) Env. Sci. Tech. 4:589
48. Goldstein IF (1976) Use of Aerometric Data to Monitor Health Effects, APCA Paper 76-32.6, presented at the 69th Annual Meeting, APCA, Portland, OR
49. Simon C (1978) Bull. NY Acad. Med. 54:1012
50. Glasser M, Greenburg L, Field F (1967) Arch. Env. Hlth. 15:684
51. Greenburg L, Field F, Ehrardt CL, Glasser M, Reed JI (1967) Arch. Env. Hlth. 15:430

52. Glasser M, Greenburg L (1971) Arch. Env. Hlth 22:334
53. Buechley RW, Riggan WB, Hasselblad V, Van Bruggen JB (1973) SO$_2$ levels and perturbations in mortality, Arch. Env. Health, 27:134
54. Buechley RW (1975) SO$_2$ levels, 1967–72 and perturbations in mortality, report to the National Institutes of Health (NIEHS). Contract NO1-ES-5-2101
55. Schimmel H, Greenburg L (1972) J. APCA 22:507
56. Schimmel H, Murawski TJ, Gutfield N (1974) Relation of Pollution to Mortality, presented at the 67th Annual Mtg. APCA, APCA Paper 74-220
57. Schimmel H, Murawski TJ (1976) J. Occ. Medicine 18:316
58. Schimmel H (1978) Bul. New York Acad. of Med. No. 54, 11:1052
59. Özkaynak H, Spengler JD, Garsd A, Thurston GD (1986) Assessment of Population Health Risks Resulting from Exposures to Airborne Particles, in: Aerosols: Research, Risk Assessment, and Control Strategies, Lee SD, Schneider T, Grant LD, Verkerk PJ (eds) Lewis Publ, Inc., Chelsea, Michigan, p 1067
60. Özkaynak H, Spengler JD (1985) Env. Health Perspectives 64:45
61. Bloomfield P, Mazumdar S, Schimmel H (1980) Regional Analysis of the Impact of Air Pollution on Human Mortality, Technical Report No. 165, Series 2, Dept. of Statistics, Princeton University
62. Pickles JH (1980) The Use of Regression Models in Epidemiological Analyses of Daily Mortality Data, Laboratory Note RD/L/N 213/79, Central Electricity Research Laboratories, Leatherhead, UK
63. Wyzga RE (1978) J. Amer. Stat. Assoc. 73:463
64. Wyzga RE (1977) Proc. Social Stat. Section, Amer. Stat. Assoc., Washington, DC, p 660
65. Mazumdar S, Sussman N (1983) Arch. Env. Health 38:17
66. Mills CA (1960) Am. J. Med. Sci. 239:81/307
67. Hexter AC, Goldsmith JR (1971) Science 172:265
68. Hatzakis A, Katsouyanni K, Kalandidi A, Day N, Trichopoulos D (1986) Int. J. Epidemiology 15:73
69. Katsouyanni K (1988) (personal communication)
70. Watanabe H, Kaneko F (1970) Excess Death Study of Air Pollution, Proc. 2nd Int. Clean Air Congress, p 199
71. Kevany J, Rooney M, Kennedy J (1975) Irish J. Med. Sci. 144:102
72. Neuberger M, Rutkowski A, Friza H, Haider M (1987) Forum-Stadt-Hygiene 38:7
73. Lazar P (1981) Geographical Correlations between Disease and Environmental Exposures, in: Perspectives in Medical Statistics: Proceedings of the Rome Symposium on Medical Statistics 1980, Bithell JF, Coppi R (eds) Academic Press, London, p 21 (Also see Breslow L, Enstrom JE (1980) Prev. Med. 9:469)
74. Cowell MJ, Hirst BL (1979) Mortality Differences Between Smokers and Non-Smokers, State Mutual Life Assurance Company of America, Worcester, MA, October 22
75. Breslow L, Enstrom JE (1980) Prev. Med. 9:469
76. Piantidosi S, Byar DP, Green SB (1988) Am. J. Epidemiology 127:893
77. Snedecor GW, Cochran WG (1967) Statistical Methods, 6th ed., Iowa State University Press, Ames, Iowa, p 164
78. Nielsen A, Clemmensen J (1954) Danish Medical Bulletin 1:194
79. Rencher AC, Pun FC (1980) Technometrics 22:49
80. Lipfert FW (1984) J. Env. Econ. & Mgmt. 11:208
81. Lave LB (1982) Quantitative Risk Assessment in Regulation, Brookings Institution, Washington, DC
82. Feinstein AR (1988) Science 242:1257
83. Mills CA, Mills-Porter M (1948) Occup. Med. 5:614
84. Mills CA (1952) Am. J. Med. Sci. 224:403
85. Manos NE, Fisher GF (1959) J. APCA 9:5
86. U.S. Dept. of Health, Education and Welfare (1979) Smoking and Health, U.S. Gov't Printing Office. DHEW Publ. No. (PHS) 79-50066
87. Schiffman R, Landau E (1961) J. APCA 11:384
88. Prindle RA (1959) J. APCA 9:12

89. (a) Winkelstein W Jr, Kantor S, Davis EW, Maneri CS, Mosher WE (1967) Arch. Env. Health 14:162; (b) (1968) Arch. Env. Health 16:401
90. Winkelstein W Jr, Gay ML (1970) Arteriosclerotic Heart Disease and Cerebrovascular Disease. Further Observations on the Relationship of Suspended Particulate Air Pollution and Mortality in the Erie County Air Pollution Study, in: Proc. Inst. Env. Sciences, 16th Annual Meeting, Boston, p 441
91. Winkelstein W Jr, Kantor S (1969) Arch. Env. Health 18:544
92. Stocks P (1960) Brit. J. Cancer 14:397
93. Grove RD, Hetzel AM (1968) Vital Statistics Rates of the United States, 1940–60, U.S. Dept. of H.E.W., PHS Publ. 1677, U.S. Gov't. Printing Office, Washington, DC
94. Finch SJ, Morris SC (1979) Consistency of Reported Health Effects of Air Pollution, in: Advances in Environmental Science and Engineering, Pfaffin JR, Ziegler EN (eds) Gordon and Breach Science Publ, London, p 106
95. Fleissner ML, Gregory A, Mosher WE (1974) The Contribution of Air Pollution to the Risk of Chronic Respiratory Disease Mortality, presented at the Annual Meeting of the Amer. Public Health Ass., New Orleans
96. Finch SJ, Novak K, Ouyang SP, Schaik JV, Longitudinal Study of Association Between Air Pollution Reduction and Changes in Mortality Ratios in Buffalo (unpublished BNL memo)
97. Zeidberg LD, Horton RJM, Landau E (1967) Arch. Env. Health 15:214
98. Zeidberg LD, Horton RJM, Landau E (1967) Arch. Env. Health 15:225
99. Hagstrom RM, Sprague HA, Landau E (1967) Arch. Env. Health 15:237
100. Hagstrom RM, Sprague HA (1969) Arch. Env. Health 18:503
101. Lave LB, Seskin EP (1978) Air pollution and human health, Johns Hopkins University Press
102. Lave LB, Seskin EP (1970) Science 169:723
103. Lipfert FW (1980) Energy Systems and Policy 3:367
104. Landau E (1978) The Nation's Health
105. Cooper DE, Hamilton WC, (1979) Amer. Mining Congress J
106. Evans JS, Kinney PL, Koehler JL, Cooper DW (1984) J. APCA 34:551
107. Chappie M, Lave LB (1982) J. Urban Econ. 12:346
108. Lipfert FW (1978) Ph.D. Dissertation, Union Graduate School, Cincinnati, Ohio (Available from Eureka Publications, Mantua, NJ 08051; portions published in: (a) (1980) Sci. Total Env. 15:103; (b) (1980) Sci. Total Env. 16:165; (c) (1980) J. APCA 30:336
109. Crocker TD, Schulze W, Ben-David S, Kneese A (1979) Methods development for assessing air pollution control benefits. Vol I, Economics of air pollution epidemiology, EPA-600/5-79-001a, U.S. Environmental Protection Agency, Washington, DC
110. Lipfert FW (1977) "The Association of Air Pollution and Human Mortality: "Regression Analysis for 136 U.S. Cities, 1969", APCA Paper 18.7, presented at the 70th APCA Annual Meeting, Toronto
111. Shannon JD (1981) Atm. Env. 15:689
112. Thibodeau LA, Reed RB, Bishop YMM, Kammerman LA (1980) Env. Hlth. Persp. 34:165
113. McDonald GC, Schwing RC (1973) Technometrics 15:463
114. Schwing RC, McDonald GC (1976) Sci. Tot. Envir. 5:139
115. Thomas TJ (1975) An Investigation into Excess Mortality and Morbidity Due to Air Pollution, Ph.D. Dissertation, Purdue University
116. Mendelsohn R, Orcutt G (1979) J. Envir. Econ. & Mgmt. 6:85
117. Orcutt GH, Franklin SD, Mendelsohn R, Smith JD (1977) Amer. Econ. Rev. 67:260
118. Lipfert FW (1983) J. Envir. Econ. & Mgmt. 10:184
119. Selvin S, Merrill D, Wong L, Sacks ST (1984) Env. Health Perspectives 54:333
120. Gerking S, Schulze W (1981) AEA Papers and Proceedings 71:228
121. Namekata T, Carnow BM, Reda DJ, O'Farrell EB, Marselle JR (1979) Model for Measuring the Health Impact from Changing Levels of Ambient Air Pollution, Report to U.S. Environmental Protection Agency, EPA-600/1-79-034
122. Özkaynak H, Thurston G (1987) Risk Analysis 7:449
123. Winkelstein W Jr (1985) Int. J. Epidemiol. 14:39
124. Gregor JJ (1977) Mortality and Air Quality, the 1968–72 Allegheny County Experience, Report

#30, Center for the Study of Environmental Policy, Pennsylvania State University, University Park, Pennsylvania

125. Jacobs CF, Langdoc BA (1972) Health Services Reports 87:623
126. Plagiannakos T, Parker J (1988) An Assessment of Air Pollution Effects on Human Health in Ontario, Report No. 706.01 (#260), Energy Economics Section, Economics and Forecast Division, Ontario Hydro, Toronto
127. Pemberton J, Goldberg C (1954) Br. Med. J. ii:567
128. Gorham E (1958) Lancet ii, 691
129. Daly C (1954) Br. Med. J. ii:687
130. Stocks P (1959) Br. Med. J. i:74
131. Gardner MJ, Crawford MD, Norris JN (1969) Brit. J. Prev. Soc. Med. 23:133
132. Chinn S, Florey CduV, Baldwin IG, Gorgol M (1981) J. Epidem. Comm. Health 35:174. Also see Florey CduV, Chinn S, Baldwin IG, Gorgol M (1980) "Final Report on Mortality and Air Pollution in the County Boroughs of England and Wales", 1969–73
133. Dean G, Lee PN, Todd GF, Wicken AJ (1978) Report on a second retrospective study in North-East England, Tobacco Research Council Research Paper 14, Parts 1 and 2, London
134. Imai M, Yoshida K, Kitabatake M (1986) Arch. Env. Health 41:29
135. Kitagawa T (1984) J. APCA 34:743
136. Imai M (personal communication to FW Lipfert, May 31, 1986)
137. Krzyzanowski M, Wojtyniak B (1982) J. Epid. Comm. Health 36:262
138. Morris SC, Shapiro MA, Waller JH (1976) Arch. Env. Health 31:248
139. Göttinger HW (1983) Env. Int. 9:207
140. Blakeslee S (1988) New York Times, March 24, p B6
141. U.S. Dept. of Health, Education, and Welfare (1974) Mortality Trends for Leading Causes of Death, United States—1950–69, DHEW Publication No. (HRA) 74-1853
142. Keys A (1980) Seven Countries, A Multivariate Analysis of Death and Coronary Heart Disease. Harvard University Press
143. Lipfert FW (1988) Estimates of Historic Urban Air Quality Trends and Precipitation Acidity in Selected U.S. Cities (1880–1980), Brookhaven National Laboratory Report to the National Park Service
144. Gover M (1949) Public Health Rep. 64:439
145. Hechter HH, Borhani NO (1965) Public Health Rep. 80:11
146. Enterline PE, Stewart WH (1956) Public Health Rep. 71:849
147. Enterline PE, Rikli AE, Sauer HI, Hyman M (1960) Public Health Rep. 75:759
148. Sauer HI, Payne GH, Council CR, Terrell JC (1966) Public Health Rep. 81:455
149. Sauer HI, Donnell HD Jr (1971) J. Gerontolgy 25:83
150. Sauer HI, Brand FR (1971) Geographic Patterns in the Risk of Dying, in Environmental Geochemistry in Health and Disease, Geological Society of America Memoir 123
151. Erhardt CL, Berlin JE (1974) Mortality and Morbidity in the United States, Harvard University Press, Cambridge, MA
152. Kitagawa EJ, Hauser PM (1973) Differential Mortality in the United States: A Study in Socioeconomic Epidemiology, Harvard University Press, Cambridge, MA
153. Doll R, Peto R (1981) J. Nat. Cancer Inst. 66:1193
154. Ware JH, Thibodeau LA, Speizer FE, Colome S, Ferris BG Jr (1981) Env. Health Persp. 41:255
155. Kleinman JC, DeGruttola VG, Cohen BB, Madans JH, (1981) J. Chron. Dis. 34:11
156. Glass GV, McGaw B, Smith ML (1981) Meta-Analysis in Social Research, Sage Publications, Beverly Hills, CA
157. Havlik RJ, Feinleib M (eds) (1979) Proc. Conf. on the Decline in Coronary Heart Disease Mortality, NIH Publ. 79–1610, U.S. Dept. of H.E.W.
158. Garland C, Barrett-Connor E, Suarez L, Criqui MH, Wingard DL (1985) Am. J. Epidem. 121:645
159. Helsing KJ, Sandler DP, Comstock GW, Chee E (1988) Am. J. Epidem. 127:915
160. Wells AJ, (1988) Envir. Int. 14:249
161. Brunekreef B, Boleii JSM (1982) Int. Arch. Occup. Envirn. Health 50:299

Subject Index

O. Hutzinger, University of Bayreuth (Ed.)

Detergents

With contributions by numerous experts

1991. Approx. 400 pp. 48 figs. 116 tabs.
(The Handbook of Environmental Chemistry, Vol. 3,
Anthropogenic Compounds, Part F)
Hardcover. In preparation. ISBN 3-540-53797-X

Contents: *P. Christophliemk, P. Gerike, M. Potokar, Düsseldorf,
FRG:* Zeolites. – *J. S. Falcone, J. G. Blumberg, Conchohocken, PA:*
Anthropogenic Silicates. – *J. L. Hamelink, Midland, MI:*
Silicones. – *D. Gleisberg, Hürth, FRG:* Phosphate. –
W. E. Gledhill, T. Feijtel, St. Louis, MO: Environmental Properties
and Safety Assessment of Organic Phosphonates Used for
Detergent and Water Treatment Application. – *H. L. Hoyt,
H. L. Gewanter, Groton, CT:* Citrate. – *H.-J. Opgenorth,
Ludwigshafen, FRG:* Polymeric Materials, Polycarboxylates. –
J. G. Batelaan, C. G. van Ginkel, F. Balk, Arnhem, NL: CMC. –
P. A. Gilbert, Merseyside, UK: EDTA and
TAED. – *R. S. Boethling, D. G. Lynch,
Washington D.C.:* Quaternary Ammo-
nium Surfactants. – *H. A. Painter,
Knebworth, UK:* Anionic Surfactants. –
*R. J. Watkinson, G. C. Mitchell, M. S. Holt,
Sittingbourne, UK:* The Environmental
Chemistry, Fate and Effects of Nonionic
Surfactants. – *K. Raymond, L. Butterwick,
London, UK:* Perborate. – *H. B. Kramer,
Amsterdam, NL:* Fluorescent Whitening
Agents.

Springer-Verlag
Berlin
Heidelberg
New York
London
Paris
Tokyo
Hong Kong
Barcelona
Budapest

O. Hutzinger,
University of Bayreuth (Ed.)

Anthropogenic Compounds

With contributions by numerous experts

1991. XI, 237 pp. 25 figs. 75 tabs.
(The Handbook of Environmental Chemistry,
Vol. 3, Part G) Hardcover DM 168,–
ISBN 3-540-53198-X

Contents: *F. Brochhagen, Bergisch-Gladbach, FRG:* Isocyanates. – *H. Fiedler, W. Mücke, Bayreuth, FRG:* Nitro Derivatives of Polycyclic Aromatic Hydrocarbons (NO$_2$-PAH). – *J. Konietzko, K. Mross, Mainz, FRG:* Chlorinated Ethanes: General Sources, Biological Effects, and Environmental Fate. – *D. Rosenblatt, E. P. Burrows, W. R. Mitchell, D. L. Parmer, Baltimore, MD:* Organic Explosives and Related Compounds.

Springer-Verlag
Berlin
Heidelberg
New York
London
Paris
Tokyo
Hong Kong
Barcelona
Budapest